For Love of the Automobile

For Love of the Automobile

Looking Back into the History of Our Desires

Wolfgang Sachs

Translated from the German by Don Reneau

UNIVERSITY OF CALIFORNIA PRESS
Berkeley · *Los Angeles* · *Oxford*

Originally published as *Die Liebe zum Automobil: ein Rückblick in die Geschichte unserer Wünsche*
© 1984 Rowohlt Verlag, Reinbek bei Hamburg

University of California Press
Berkeley and Los Angeles, California

University of California Press, Ltd.
Oxford, England

© 1992 by
The Regents of the University of California

Library of Congress Cataloging-in-Publication Data
Sachs, Wolfgang.
  [Liebe zum Automobil. English]
  For love of the automobile : looking back into the history of our desires /
Wolfgang Sachs ; translated from the German by Don Reneau
      p.   cm.
  Translation of : Die Liebe zum Automobil.
  Includes bibliographical references.
  ISBN 0-520-06878-5
  1. Automobiles—Social aspects—History.   2. Automobiles—
History.   3. Technology and civilization.   I. Title.
HE5613.S2313   1992
303.48'32—dc20   91-8902
                                                        CIP

Printed in the United States of America

The paper used in this publication meets the minimum requirements
of American National Standard for Information Sciences—Perma-
nence of Paper for Printed Library Materials, ANSI Z39.48-1984. ∞

Contents

DISENCHANTMENT

PROSPECTS

Preface

There is no shortage of writings critical of the automobile. For that reason such topics as carbon monoxide and decibel ranges, dying forests, lung cancer, and accident casualties will be of marginal concern in this book. Nevertheless, the charges do not seem to be causing great disquiet among automobile enthusiasts. The problem with the automobile today consists precisely in the fact that the automobile is *not* a problem. Why, I asked myself, does the loyalty to automobiles remain so unassailable, even though everyone knows that cars already have their future behind them?

Love only rarely listens to reason. So it seems here as well. At stake are our needs and preferences—and they have proved astoundingly resistant to the clever arguments and cunning mathematics of the critics of the automobile. That is why I have undertaken to study technological development as a chapter in the history of mentalities, to measure how much latitude we are left by our need for an ecological future.

The automobile is much more than a mere means of transportation; rather, it is wholly imbued with feelings and desires that raise it to the level of a cultural symbol. Behind the gradual infiltration of the automobile into the world of our dreams lie many stories: ones of disdain for the unmendable horse, of female coquetry, of the driver's megalomania, of the sense of having a miracle parked in the drive, and of the generalized desire for social betterment. A technological history, setting one type of automobile next to others and singing a devotional hymn to increasing perfection, is blind to human needs and cultural significations; it fails to consider that every technology is the product of a historical period, in which it rises to prominence and then disappears. This book, therefore, is an invitation on a journey back to the beginnings of our automotive needs, to where the breast first swelled with the pride of independence, where the love of speed was born, where the feeling of comfort took root, and where the automobile became allied with the clock as a "time-saving machine."

From today's perspective, the story is not a triumphal one, replete with flags and fanfare. The history of the automobile can ultimately be read as a morality play about the withering of a historical project. The dreams are aging in our day: boredom with motorization is widespread, and contrary images are becoming evident; the preference for bicycles is growing, and the idea of an unhurried society finds fertile soil on which to fall. Today it is the personal computer, more than the internal combustion engine, that causes excitement. Is the microchip not for our children what the engine was for our grandfathers? Thus it appears today, nearly one hundred years after Carl Benz first rattled through the streets of Mannheim, that a eulogy is in order, a eulogy to the history of the excitement caused by the automobile.

For suggestions and encouragement, for criticism and support, I want to thank Ingke Brodersen, Bernhard Fliß, Christian Holz, Helmut Holzapfel, Ivan Illich, Ursula Juch-Neubauer, Jobst Kraus, Jean Robert, Klaus Traube, Otto Ullrich, and Thomas Weymar. I have profited especially from years of discussions with my friends and colleagues from the "Energie und Gesellschaft" (Energy and society) project at the Technische Universität in Berlin. Their study, *Szenario Auto 2000. Wege zu einem ökologisch und sozial veträglichen Autoverkehr* (Auto scenario 2000: Toward an ecologically and socially tolerable traffic system), can be read as the traffic planners' counterpart to this cultural history.

Wolfgang Sachs
Berlin, May 1984

Stations

When at some later time cultural historians try to label the epoch extending from the beginning of the nineteenth to after the turn of the twentieth century, they will best designate it the epoch of great advances in transportation, of great progress in vehicular technology. At the dawn of this epoch comes the invention of the railroad, the locomotive, which for the first time since the beginning of cultural history replaced the horse in its dominant role as the most important means of travel and transport on land . . . ; and at the close of the same epoch the automobile begins its triumphal march. . . . Thus does the automobile appear to complete what the railroad began and the age of the horse to come to an end, finished by the power of mechanized natural forces, which are destined to become the driving element in travel and transport forever after.

> Theo Wolff, *Vom Ochsenwagen zum Automobil.*
> *Geschichte der Wagenfahrzeuge und des Fahrwesens*
> *von ältester bis neuester Zeit* (From ox-cart to
> automobile: History of vehicular traffic from earliest
> to most recent times) (Leipzig, 1909), 159f.

Pleasures for the Wealthy
(1890–1914)

The sober journalistic tone of the *Vorarlberger Landeszeitung* barely concealed wonder as it noted on March 11, 1893, the first appearance of a "Patent-Autocar Benz" on the streets of Bregenz, Austria:

> A carriage with a gasoline engine has been here for a few days, in the private possession of Mr. Eugen v. Zardetti. In its construction the vehicle resembles an elegant chaise; in front, however, the car rests on a single wheel, which can be turned left or right as on a velocipede for purposes of directing the craft. Ignition of the gasoline engine is accomplished by electricity. The forward progress of the car is even, gentle, and can be raised to a very high speed. This novelty exercises a peculiar attraction: no horses are needed, no skittishness, no harness, etc.—advantages which will be of great importance only once such vehicles become cheaper.[1]

In the skeptical astonishment of this early observer flashes an idea that will continue to define much of the "essence" of the automobile for succeeding generations: the liberation of speed from the fetters of corporeal nature. The coach rolls to its destination entirely without horses; therein lies the novelty's "peculiar attraction." Until now the horse and carriage offered the most refined means of transport through

---

1. Quoted in H. Seper, *Damals als die Pferde scheuten. Die Geschichte der österreichischen Kraftfahrt* (Back when the horses shied: The history of Austrian motor travel) (Vienna, 1968), 17f.

3

"New! Practical! Gas-powered motorcar, patented in all industrial countries. Entirely supersedes the horse and wagon . . . Always starts right up . . . very low operating costs . . . comfortable and totally safe! Patented motorcar with removable title and splash-leather." From the first automobile prospectus, 1888.

Hygienic Considerations. Driving in a motorcar, like all mechanical calisthenics, effects a brisk activization of the entire organism, but, in comparison with other calisthenic methods, possesses notable qualities of its own. In contrast to indoor calisthenics, the flow of fresh air in particular must be considered, which stimulates the activities of the skin and lungs in a pleasant manner and, in so doing, initiates an extremely advantageous unburdening of the internal organs, which are quite excessively gorged with blood. Horseback riding seems to many persons too rigorous, and driving in a normal wagon without air-filled tires too rough; in contrast, driving in a motorcar consists of a light and gentle floating motion, which makes itself felt in a pleasant manner, rather like riding in a skiff on calm water. With the vehicle's low center of gravity, the rough bumps of the road are almost entirely absorbed by the pneumatics and the springs, so that there is no feeling of inflexibility and stiffness after a long drive, as one often has stepping out of a normal carriage or railroad car. Rather, one has the feeling of being pleasantly tired, much as after a breezy climb out of doors one feels intensified sleepiness and hunger. A highly advantageous effect on the nerves goes hand in hand with the beneficial relaxation caused by scenic landscapes and the unburdening of the internal organs. To be sure, several prerequisites must be fulfilled: neither dawdle along your way, nor race, but proceed rather at a moderate tempo, and, indeed, systematically, mornings and afternoons, in summer and winter, outfitted when necessary with glasses, leather gloves, a fur wrap, etc. As a result of the beneficial effect on the nerves, we find motorcar enthusiasts particularly among those devoted to the performance of mental labor.

From the entry entitled "Motorwagen," *Meyers Großes Konversationslexikon*, 6th ed., vol. 14 (1909)

the streets, and although bicycles were seen in isolated instances, most people had to be content with "the cobbler's conveyance"—that is, going on foot. The horse and carriage had insured that more genteel travelers were protected from the muddy streets, raised above the many, and, drawn by an impersonal force, able to go about their business. Did it make sense to release the horse from the harness and move to motorcars? That was the burning question in refined circles.

The wait for an answer from an expert source was not long in coming. L. Baundry de Saunier, a French authority in "automobilism" who was also recognized in Germany, made short work of the old ways in the first popular book on the automobile: "It can . . . be said with certainty that the horse—a weak, dangerous, costly, and dirty motor, easily broken and hard to repair—. . . is destined to disappear."[2] With evident pleasure, Saunier enumerated the organic frailties of the horse. Not only was it sluggish and prone to exhaustion; beyond that it could not be repaired, for "its bones cannot be soldered, and once the front piston rods, its knees, are cracked, one cannot return them even superficially to order by applying some sort of enamel." Then, too, the "oat motor" is vulnerable: "If the oats with which it is fed contain dust, the air intake sustains immediate damage, and it coughs; should the water it takes in be too cold, then its drainage ports contract, often entangling themselves in the most terrible fashion." The fact that the horse lives becomes its doom; a motor neither sleeps nor falls ill, nor does it leave piles of steaming waste in its trail or unexpectedly bolt. And, last but not least, "the motor run on oats possesses one flaw that reason, from the perspective of economy, must characterize as nearly monstrous: to continue to consume even when it is performing no work." Not without melancholy, but in the spirit of the time, the verdict falls: the horse is frail, whereas the engine is tireless.

Masters of Time and Space

It was certainly easier to imagine the superiority of the motor than to count on it. To be a "motorist," as the racy sport and pleasure drivers of the cities soon came to be called, one needed more than money; muscle and courage were required as well. Automobiles resembled untamed animals in the early years, with sudden swings of mood and a tendency to dangerous reactions. It took a strong arm to work the crank (which would likely kick back) and so coax a sound from the monster's insides, until it began to roar and shudder and belch its stinking vapors.

Driving was still an adventure, producing above all the gratification of having successfully overcome the fear. But then there was the fear's reward: breath-taking speed. To charge across the land, manning the helm with skill and stamina and utter presence of mind, leaving onlookers behind in the dust from the road—who could still believe that happiness rode on the back of a horse?

The driver was first of all a sportsman taking up the reins of technology, and races made the automobile the topic of the day. Beginning in 1895 with a Paris-Bordeaux-Paris race, new long-distance runs,

2. L. Baundry de Saunier, *Grundbegriffe des Automobilismus* (Fundamental principles of automobilism) (Vienna, Leipzig, 1902), 7.

> A Daimler is a worthy beast,
> Pulls like an ox, seen west and east.
> It doesn't feed while in the stall,
> And only drinks when work's the call.
> It threshes, saws, and stands your loan,
> When you fall short, a common moan.
> Ne'er ill in foot and mouth and bite,
> And never up and goes on strike.
> It doesn't scorn, attack with horn,
> Does not consume your hard-grown corn.
> So buy yourself just one such beast,
> Forevermore you'll lack the least.
>
> Advertisement verse, Cannstätt Fair, 1897.

races, and rallies were added to the calendar every year. As early as 1898, when the winner of the first German motor contest, from Berlin to Potsdam and back, achieved an average speed of 25.6 kilometers per hour, the races proved to a society eager to embrace speed the superiority and reliability of the "horseless carriage." It was the dawn of a time of ever-increasing speed (in just less than a decade the maximum jumped from 25 km/h to 100 km/h)—exciting for those who could make it their own, threatening to those forced to watch from the sidelines.

Mechanized speed and power created a sharp perceptual contrast between the age of the automobile and the era of the horse, and between the experience of individual mobility and reliance on the railroad. The horse and carriage, the traditional insignia of privilege, had declined in rank over the course of the nineteenth century, to the point that whenever a train overtook a coach, the rail passengers would laugh sneeringly out the windows. So ladies and gentlemen of a better sort had to condescend to train travel as well; forced by necessity, they became inmates of mass transportation. Technically and organizationally, the railroad was a thoroughly postfeudal means of conveyance. What the carriage had bestowed in the way of freedom and prestige now fell beneath the wheels of a steam engine racing to its scheduled destination on unmovable tracks.

"Freedom has been sacrificed to speed. The train ticket"—so ran the lament of Otto Julius Bierbaum, the Carl Benz of automobile interpretation, in 1902—

is purchased not only with money, but also with the forfeiture of one's right of self-determination for a certain while. Whoever goes traveling in a railway coach forgoes, for a time, his freedom. Every trip made by railway is also a transport of prisoners; the wardens are called conduc-

From the horseless carriage . . .

tors, which does not, however, always obligate them to courtesy; the
prison rules are called railway regulations . . . ; [and] as the solitary
cell system is too dear, the prisoners are transported . . . together in
groups, more or less large, whereby, admittedly, some consideration is
shown for the capacity of one's purse.[3]

Bierbaum feels justified in indulging some strong words, a proof of
how greatly the refined world felt its self-respect to be threatened by
the railroad. What the common people welcomed as a democratic advance, individuals of more privileged position greeted with a snort. The
railroad (like, incidentally, its contemporary, the department store) ignored differences of social status and deprived the cultivated lifestyle
of its very basis. Where everyone, whether rich or poor, is reduced to a
piece of freight, pride of place no longer finds nourishment. The epoch
of the equipage had apparently come to an end.

Against this background, the automobile accrued what one could
call a restorative significance. The ideal of the era of the carriage could
arise anew with none of the frailties of organic horsepower—a mix
of restoration and progress that opened the purses of the well-heeled.
The writer Julius Otto Bierbaum lent words to this sense of a curiously
forward-looking nostalgia in the first German-language book devoted
to an automobile trip: "The meaning of the automobile is freedom,
self-possession, self-discipline, and ease. In it the traveling coach is re-
vived in all its poetic plenitude, but in a form endlessly enriched by the
former's exquisite potential for intensified and simultaneously expanded gratification."[4]

3. Otto Julius Bierbaum, *Yankeedoodlefahrt und andere Reisegeschichten* (A yan-
kee doodle trip and other travel stories) (Munich, 1910), 467f.
4. Otto Julius Bierbaum, *Eine empfindsame Reise im Automobil. Von Berlin nach
Sorrent und zurück an den Rhein* (A sentimental journey by automobile: From Berlin to
Sorrento and back to the Rhine) (Munich, 1903), 268f.

. . . to the luxury sedan. (Courtesy Daimler-Benz Museum, Stuttgart-Untertürkheim)

No more missed trains, no more overflowing compartments, no more predetermined routes; in its commentary on the change, the *Allgemeine Automobil-Zeitung* of 1906 waxed philosophical:

> The automobile: it will grant to human beings their conquest over time and space by virtue of its speed of forward motion. The entire over-grown apparatus of the railroad—the network of tracks, the train stations, signal stands, supervisory personnel, and administration—falls away, and in relative freedom humanity has conquered time and space.[5]

A combination of the advantages of the carriage with those of the railway; freedom, through the power of the motor, from the exhaustible nature of the horse; and the individual mobility of the carriage—in these improvements were found the meaning of that "motorized carriage," the automobile.

It is therefore no wonder that the first generation of motorcars resembled horse-drawn carriages with a motor attached. It was as if the horses had simply been unharnessed from the shafts, in the place of which a pivoted axle was installed; the motor was then loaded into the "luggage compartment" to drive the rear axle by means of either a belt or a chain. Likewise, the coachman was retrained as a chauffeur, working now with a steering column instead of reins. As far as first appearances were concerned, the motorcar directly inherited the symbolic features of the coach, even if those who dared to drive the still uncanny vehicles came from the ranks of well-to-do inventors and technology enthusiasts.

Automobile design began to free itself from the model of the coach only after the turn of the century. The motor made its way to the front,

5. *Allgemeine Automobil-Zeitung*, 1906, no. 17, 33.

where it was encased in a tin housing. Seating was no longer arranged so that passengers faced one another. The wheelbase became longer, and a redesigned body provided both the engine compartment and the driver and passenger area with a belt-high enclosure. But these changes did not yet satisfy the aesthetic requirements of Bierbaum, who, in his *Automobilreise* (A sentimental journey by automobile, from Berlin to Sorrento and back to the Rhine), pressed on the designers the following advice:

> The aesthetic of the automobile lingers in its infancy. One might say that its beauty suffers at present because the designers have not yet completely forgotten the horse—namely, the horse before the wagon. . . . [Automobiles] look like draught wagons without a drive team. An automated wagon ought to have sufficient self-confidence to look like a machine. And it can be pretty. I choose not to say, "pretty as a horse." Of such a thing of beauty only our dear God is capable. . . . The form of the automobile must grow organically out of its mechanics and chassis, but must appeal down to the tiniest of its curves to the aesthetic sense and, in addition, be practical and comfortable.[6]

With near precision, Bierbaum's proposals anticipated the automobile design of the 1920s, when, influenced by the Deutscher Werkbund, a search was mounted for the beauty of technologically determined form.

For the time being, however, design reflected the adoption of the automobile by the aristocracy and upper bourgeoisie, who used its speed and power to display their social superiority. It had always been a sign of power to master an extensive spatial range and leave others staring in one's wake. In addition, a transportation vehicle served better than any other object as a status symbol, because by its very nature it commanded public attention: one drove on the streets in clear view of everyone. An elongated hood expressed social power in addition to mechanical force; and cars with closed bodies, which appeared more and more frequently after 1910, were reminiscent of equipages, where the driver sat in the rain while the more refined passengers rode in agreeable comfort within. Beyond all technical requirements, design underlined this social significance of the automobile. In accord with its presentation as a "mobile money temple," the design of the front and rear of the vehicle took over aesthetically: there the automobile's formal technical requirements could be exceeded ostentatiously by means of artistic and architectural motifs.[7] Temple themes inspired radiator

6. Bierbaum, *Eine empfindsame Reise*, 278f.
7. Joachim Petsch, *Geschichte des Auto-Designs* (Cologne: Du Mont, 1982), 18.

Automobile Club: "If I may, Board Director Fränkel, forty horsepower."
"A pleasure, Count Dohna, sixty HP." Caricature by Bruno Paul in
Simplicissimus, *1903.*

grills, giving the visible side of the radiator the significance of a facade. As in classical architecture, geometric precision and the tension between horizontal and vertical lines determined the contours of the body, while the *Jugendstil* preference for sweeping lines and the ornamental was seen in some exteriors. This formal syntax, applied to the vehicle rather than derived organically from it, expressed in traditional motifs the message that did not emerge from formal design alone: the car is a luxury item.

The upper elites thereby enlisted the car into the ranks of their objects of self-representation, just as, conversely, the automobile's own artistic pretensions graced its owner with higher social status. The automobile caused no immediate revolution in mobility, but it did in the dominant symbols of prestige. In this essential aspect it was welcomed by the doctors and lawyers, the entrepreneurs and the upper middle class, who used the car to demonstrate their elevated social position, though they were not of noble lineage. Right around the turn of the century these bourgeois groups, especially in the cities, were gaining in importance, while the power of the landed aristocracy was increasingly in decline. For these nouveaux riches the automobile arrived just in time, allowing them to enter the scene as masters of the dawning epoch. The automobile took up and incorporated the interests and ideals of the rising, moneyed, urban bourgeoisie. With the aim of demonstrating that the bearers of these ideals and interests had the say in society, the bourgeoisie arrogated to itself aristrocratic symbols, using the proper insignia to lend emphasis and legitimacy to its claim. The masters of time and space demanded that they also be the new masters of the social order. And, reversing the logic, the new masters of the social order documented their claim through their new power over time and space.

Who Owns the Road?

> Never in my life have I been cursed at so frequently as on my automobile trip in the year 1902. Every German dialect from Berlin through Dresden, Vienna, and Munich to Bolzano was represented, as well as all the idioms of the Italian language . . .—not to mention all the wordless curses: shaking fists, stuck-out tongues, bared behinds, and others besides.[8]

As pleasant as Mr. Bierbaum's automobile trip with his wife may have been, he could not refrain from discussing its dark sides: one man's pleasure is another man's pain. Curses and shaking fists, flying stones, journalistic libel, and attempted acts of parliament accompanied the motorcar on its course all the way through the first decade of this century. The alluring mastery over time and space had to be achieved over the wrath of the people.

The reason is simple. The automobile alone does not suffice but, to be driven out and about, is greatly in need of streets. A platitude, indeed, but one rich in consequences. For in this truth lies the proof that the automobile changed the face of our society. While other mechanical novelties, like a typewriter or a vacuum cleaner, make no impositions beyond their owner's four walls, the car requires free passage in

8. Ibid., 285.

To the Honorable Chief of Police,

Yesterday evening at six o'clock on the rue de Courcelles, I, my wife, and my children were nearly run over by a gentleman in an automobile who came racing by at the speed of a train.

It was of course impossible to catch him. When I turned to a policeman to inquire whether the gentleman lived in the neighborhood and whether we might have any chance of finding him, I was told: "O God, my good sir, we are powerless when it comes to these people. As you know, they can get away from us by fleeing."

My honorable Police Chief, it is imperative that you obligate these speeders to mark their automobiles with a visible number, which would help one track them down after they have fled. And this requirement must be put in effect tomorrow, not six months from now.

Until it has been, I must count myself among those who believe there is no safety anymore on the streets of Paris. And since your policemen declare themselves to be powerless, let me have the honor of informing you that from this day on I shall carry a revolver in my pocket whenever I go out, and I will shoot at the next crazy idiot who tries to flee after he was on the verge of running over my family and me.

Open Letter to the Chief of Police, *Le Journal*, 1896

the streets and demands also of those without a car that they behave according to the rules of its existence. Thus the automobile was from the start not only a technical problem, but a problem that concerned the streets and conventional conduct. A history of the automobile must be equally a history of the environment and behavior.

The conquest of the streets was bound to meet with objection; they were, after all, inhabited by pedestrians, horse-drawn conveyances of various types, children at play, and all kinds of fowl. The noise of protest pressed all the way into the measured world of parliament, as evidenced, for example, by the automobile bill brought before the Prussian legislature in January 1908. The representative Count Cramer expounded in its defense. While his friends did not want to put any obstacles in the way of the automobile movement, he said, surely the latitude extended to wild motorists must have limits:

In the last six months, 2,920 automobile accidents have occurred here, and 33 people have been killed. *(Hear, hear!)* And indeed, it was primarily luxury automobiles that caused these accidents—sports cars and pleasure vehicles. The motorists often displayed an unheard-of rudeness *(Quite right!)* in rushing through at great speeds. These gentlemen place too high a value on their time. The country streets in particular are endangered, and the rural populace is embittered over the motorists to a most extreme degree, all the more so as the latter see their way clear, when once they have caused something to happen, to remove themselves from responsibility through flight.[9]

Conservative opinion in particular, with guaranteed appeal to a general ire against city dwellers' presumptuousness, condemned the car—meaning the upstart bourgeoisie. Inhabitants of the countryside were especially agitated; they felt as if a carpet of clatter, stink, and dust were being drawn over them. And this ignores the constant problem of shying horses, which, frightened by the motorized monsters, frequently bolted and overturned the cart, not only on top of the load of potatoes but on the buried driver as well. Naturally, problems got properly going only after dark! Blinded by the headlights, a horse might take to its heels; chickens, invisible at night, were fair game in the roadway; and people, cursing the power that roared down upon them, had no recourse but to leap from the road and into ditches. What delight the drivers, sitting arrogantly behind the windshield, took in the damages is evident from a journal entry of 1906 by Rudolf Diesel on his first trip:

> What a dust storm we stirred up leaving Italy! In my whole life I've never experienced such a thing again. Powdery limestone dust lay five centimeters thick on the street. Georg raced along, demanding from the car all it had to give, through the Piave Valley, and behind us there swelled a colossal cone. This white cone rose in the air and expanded to infinite proportions. The entire Piave Valley was thick with the fog, a white cloud lying over the valley all the way to the mountain ridge. We outraged the pedestrians with a gas attack—their faces pulled into a single grimace—and we left them behind in a world without definition, in which the fields and the trees in the distance had lost all color to a dry layer of powder.[10]

No wonder, given this self-righteous arrogance, that people's tempers flared, the more so since the village inhabitants themselves had to pay for damage to the streets and commons. Little wonder, too, that the rage blended with class hatred, for those who raced along the country and village streets, who drove speedily off leaving the peasants with the mess, were indeed those nouveaux riches from the cities.

Alongside the peasants and provincial authorities, individuals in-

9. *Allgemeine Automobil-Zeitung,* 1908, no. 5, 33.
10. Eugen Diesel, *Autoreise 1905* (Automobile trip 1905) (Leipzig, 1941), 190f.

Not the only ones on the road. Photograph by Jacques H. Lartique, 1908. From Lartique, Photo-Tagebuch unseres Jahrhunderts *(Lucerne: C. J. Bucher, 1970).*

volved in livery and coach fabrication and the owners of wagons opposed the spread of the automobile as well, for reasons of immediate economic interest. Only rarely did a targeted critique of progress in principle find a voice. In 1912, Dr. Michael Freiherr von Pidoll of Vienna published a "Call to Protest," in which he claimed for the public in general a right to the street:

> The alleged "street incompetence" of the public dates only from the rise of automobilism. Does this mean that the entire population has been enlisted in the service of the latter? . . . Where does the motorist get the right "to master"—as he boasts—the street? It in no way belongs to him, but to the population as a whole. Whence his right to hound the people's footsteps and dictate to them a behavior that he is justified in demanding only on his own private paths? The public street is not meant for express traffic; it belongs to the milieu of the city. . . . Should, perhaps, the public streets be kept "free of people"?[11]

11. Michael Freiherr von Pidoll, *Der heutige Automobilismus. Ein Protest und Weckruf* (Automobilism today: A call to protest) (Vienna, 1912), 36ff.

Among motorists it is common to attempt to pass shying horses as quickly as possible, so to minimize the noise made by the engine in bringing the automobile to a stop, which is frightening to the animals. This point of view, however, has not been approved by the court in cases of accidents occurring upon continued driving, even accidents involving automobiles previously brought to a stop.

When, on November 17, 1902, the estate owner Mr. von G. of L. (West Prussia) passed the crest of a hill on the main road between Neuenburg and Schwetz and proceeded down the incline, a cart belonging to the blacksmith K. of Neuendorf, which, in addition to K., was carrying the latter's daughter and a coffin with a corpse inside, was approaching from the other direction. G. saw that the horse was shying, and he stopped the automobile as a result of the wagon driver's warning, "Slowly, slowly, Baron, the horse is shying." The cart belonging to K. also came to a stop, and the two vehicles stood facing each other at a distance of some three to four meters. G. now believed he could easily pass, and drove on. In doing so, he caused the rear wheels of his vehicle to collide with the cart, which the horse had meanwhile pushed backward into the middle of the road, and the driver of the wagon was thrown to the ground, resulting in an injury to his wrist. On the basis of paragraph 823, sec. I, B.G.B, the latter succeeded in receiving damages from the obstructing motorist.

The provincial court in Graudenz found these instances of negligence on the part of the defendant to pertain. For one thing, he had been driving at a high speed and was obligated, once he had brought the automobile to a stop, not to drive farther until he was certain that the vehicle could pass. In the words of the judgment: "It would therefore have been easy for the wagon driver to have jumped down, to have taken the horse, which was pushing the wagon backward, by the head, thereby conducting the vehicle securely past the stopped automobile. This possibility must have been recognized by the defendant, who is the owner of an estate and a cavalry officer. It may further be correct that the defendant was justified in hoping to be able to pass the plaintiff's wagon, but so must he also acknowledge that it might just as well not have been the case. *He is not justified in deciding on a course of action that requires that his judgment necessarily be correct, where he had also to recognize that it might easily be wrong.*

(Cf. "Entsch. d. R.G.," *Juristische Wochenschrift* 56 [1905]: 154ff.) The court finds that the severe collision with the wagon is to be laid to the responsibility of the defendant, otherwise he could still have stopped his motor vehicle at the last moment. The alleged responsibility on the part of the plaintiff, for not having expertly handled his horse, which he should meanwhile have driven forward, was denied. Thus did the provincial court arrive at a *conviction* of the plaintiff.

This decision was upheld on appeal by the defendant to the superior court in Marienwerder. The superior court found further that the defendant violated paragraphs 29, sec. 2; 31, sec. 1; and 32 of the civil regulations of West Prussia, in that, confronted by a shying horse, he did not stop or did not stop as long as the avoidance of an accident and the elimination of danger required. The argument of the defendant, that he drove on in order to spare the horse the disturbing noise of the engine, was denied on the basis of expert testimony that he, in no time and with little effort, could have shut off the noise of the motor.

On the defendant's negligence according to paragraph 823, sec. I, B.G.B., the superior court determined in addition: "In fact . . . the collision was ultimately the result of the shying horse having backed up, thereby pushing the wagon behind him crossways to the road; of the moving automobile not being able to get by without colliding with the wagon or, as the expert witness V. finds, scraping it. That the defendant, as both experts find, was not prepared for the horse backing up or the wagon being pushed crossways to the road, that he, therefore, did not consider this possibility but drove on in the expectation that he would get happily by the plaintiff's cart, evidences a lack of the appropriate consideration made necessary by the exercise of conscientiousness in traffic. Had he however been aware of this possibility, then an even greater lack in the appropriate caution made necessary by the exercise of conscientiousness becomes apparent in the fact that he drove on. The circumstance that the automobile, contrary to the finding of the provincial court, is not broader in the back than in the front, provides no basis for a revision of judgment. That the horse and the plaintiff's cart could at least have been led past the stopped automobile, if the noises from the automobile that unsettled the horse had been shut off, does not appear doubtful and has not in the opinion of the court been denied by the defendant. For these reasons, the defendant's appeal was denied.

Allgemeine Automobil-Zeitung, 1906, no. 46.

THE PROHIBITION ON AUTOMOBILES IN
GRAUBÜNDEN, 1900–1925

The Swiss canton of Graubünden seemed to motorists of the time the very refuge of backwardness. Those stubborn Swiss wanted literally to ban progress and have nothing to do with automobile traffic! The columns of the European automotive press were filled with fire and fury, and readers were even exhorted to boycott Switzerland.

And, in fact, the "automobile question" was a bone of political contention in Graubünden for twenty-five years. After complaints about bolted horses, peasants driven into roadside ditches, and tourists suffering accidents had accumulated, the local council pronounced on August 8, 1900: "The driving of automobiles on all streets of the canton of Graubünden is prohibited." The administration probably did not anticipate the results of its action, in that the prohibition set the fuse for a conflict that was to flare up repeatedly. It was a conflict featuring endless debates, ceaseless efforts to obtain driving permits, a flood of pamphlets and polemics, and no less than ten referenda, until finally in 1925 the automobile was admitted to the canton streets.

The *Zurich Post* was balanced in its commentary:

> The prohibition will attract attention, perhaps provoke ridicule of the attempt to prohibit driving by police power. Nevertheless, a just view of the situation requires consideration of the particular conditions in the confederated state. . . . The streets are generally narrow and often border on a precipice. It takes only one spooked horse to cause a catastrophe. If all automobile drivers were reasonable enough to drive cautiously and at a speed suited to the conditions of the streets, the danger would not be great and one could count on the streets quickly adapting to the new vehicle, as has long been the case even on the busiest streets in Paris. The sport of the automobile, however, is still in its adolescence, and drivers speed off without consideration of others, who do value their lives and health. It appears that the local police were impotent in the face of the excesses of such drivers, such that the government decided simply to prohibit automobile traffic.

The *Bündner Tagblatt,* in contrast, struck a more drastic tone: "Peasants, herdsmen, and livestock owners and traders, after all, still live in Graubünden, and they will hardly permit the stink wagons to drive them off the streets. Or should they resort to self-defense? (We would advise them to immediately, even if at the cost of violence.—The editor.)"

Not that the prohibition brought peace to the mountaintops. A clamor for special permits set in, until the prohibition was as shot through with holes as the local cheese. This, in turn, got the irritated people involved, and, under the aegis of the confederacy's sense of freedom, a petition campaign forced a popular referendum in 1910. Wrath against the "dangerous toys of the idle sport enthusiasts" rang through the valleys, and the government was left no choice but to bet on rational far-sightedness and issue warnings against "throwing a monkey wrench into the wheels of progress." Before the vote, the following appeared in the pages of the *Neue Bündner Zeitung:*

> Two fronts have formed in the struggle: the will of the people and the will of the government. The people do not want the automobile; the government, or a part of it, wants to admit it under whatever conditions necessary. . . . When, for example, we see how our otherwise valued minister of the interior resists so bitterly, so stubbornly the will of the people, how he puts his own will in the forefront and in the middle of the issue, then we have to ask ourselves: are we living in our old confederated democracy? or have we been unconditionally delivered up to big and little kings?

Versifiers, too, lent their voices to the people:

> Will confederation's proud old right
> Of passage free even to the farmhand's sight
> By masters' lust be robbed?
> Whether masters' lust or peasants' right
> Will o'er us all exert its might
> Tomorrow will be decided.
> So rise you peasant commonweal,
> Rush to the urn with ringing peal,
> See freedom's good defended.

And the peasants apparently did rush: the prohibition on automobiles won impressive confirmation, 11,977 votes against 3,453.

In the following decade, the wind appeared to be slowly shifting. Proponents of the automobile called on "reason"—Graubünden could hardly, after all, build a "Chinese wall" around itself—and besides, the vehicles had undergone technical improvements and drivers were more responsible. Moreover, hotel owners worried about their best-paying guests, and the post office and fire department underscored the benefits of the automobile for the common good. Nonetheless, reservations concerning the "parasites in their luxury automobiles" were still sufficiently strong that even between 1920 and 1924 several referenda to permit automobile traffic were defeated. Only in 1922, when the central administration in Bern intervened and ordered the opening of streets to through traffic, was the way cleared, so that, in 1925, a scant majority, 11,318 to 10,271, decided to concede the right of way to the automobile in Graubünden, at least on the main streets.

All quotes are from Felici Maissen, *Der Kampf um das Automobil in Graubünden 1900–1925* (The fight over the automobile in Graubünden 1900–1925) (Chur: Automobilclub der Schweiz, 1968).

PEOPLE OF GRAUBÜNDEN WAKE UP!

Nine years ago you forcefully refused to allow your free land of the Alps to be defiled into a playground for automobiles, and now you're expected to declare that it was all a mistake! Next summer the **automobile** is supposed to be driving along your streets. Is that what you want? We hope the answer is:

No!

The doctor's car, the ambulance, the fire department car, the post office car—they can come and we welcome them.

Transport vehicles should be allowed in areas with streets suited to them; but they **must not become competition to the suffering Rhätische railway.** The transport of goods by automobile instead of by railway reduces the railway's receipts; no one needs an expert study to recognize that. Is the citizenry to make up the reduced receipts for the railway through taxes?

No!

Many of our streets are too narrow for any significant automobile traffic. Above all, they are not sturdy enough, so that in spring and after rains the roadbed is torn up and damaged. That raises maintenance costs. Our communities already have enough burdens to bear: for that reason transport vehicles should not be allowed to drive on streets that are too weak for them, and the useless luxury automobiles should keep off our land. Or do you, people of Graubünden, want to do compulsory service on your streets for those who, with **arrogant disdain**, race by you in their power vehicles? We think:

No!

The hotel owners call the touring automobile, or, better, the luxury automobile, their salvation. And we regret to have to take a position here against the supposed interests of the hotel owners. But we hear from the lowlands that all property and buildings located next to streets traveled by automobiles decline sharply in value. Are the many little people dwelling on our streets supposed to make this sacrifice for the hotel owners? Should the little person suffer so that the grand might **perhaps** get what they want? Is that what you desire, people of Graubünden? We hope:

No!

Today a small crack is to appear in the dam that protects us all. But that is only the beginning. Today it is supposed to be a couple of streets, tomorrow it will be all of them. They are trying to divide us, to make us easier to conquer. **Therefore, people of Graubünden, wake up; no one stays at home; every vote counts:**

No!

County Council Committee
[Appeal before the referendum of 1920 in Graubünden]

Bündner Volk, wach auf!

Vor neun Jahren hast du dich mit gewaltiger Wucht geweigert, dein freies Alpenland zum Tummelplatz des Autos erniedrigen zu lassen, und jetzt sollst du erklären, dass du damals geirrt habest. Im nächsten Sommer soll das **Auto** auf deinen Strassen fahren. Willst du das? Wir hoffen:

Nein!

Das Arztauto, das Krankenauto, das Feuerwehrauto, das Postauto mag kommen, wir begrüssen es.

Das Lastauto soll Gegenden, deren Straßen dafür geeignet sind, gestattet werden; aber es **darf keine Konkurrenz für die notleidende Rhätische Bahn** werden. Der Transport der Waren mit dem Auto statt mit der Bahn vermindert die Einnahmen der Bahn; um dies zu erkennen, braucht es kein Gutachten. Soll dann das Volk für die verminderten Einnahmen der Bahn auf dem Steuerwege aufkommen?

Nein!

Viele unserer Strassen sind für einen Autoverkehr von irgendwelcher Bedeutung zu schmal und vor allem zu leicht gebaut, so dass der Strassenkörper im Frühling und bei Regenwetter rasch aufgerissen und beschädigt wird. Das erhöht die Unterhaltungskosten. Unsere Gemeinden aber haben jetzt schon genug Lasten zu tragen: darum sollen die Lastautos auf den Strassen, die für sie zu schwach gebaut sind, nicht verkehren dürfen, und die unnützen Luxusautos sollen unserm Lande fernbleiben. Oder willst du, Bündnervolk, auf deinen Strassen Frondienste leisten für die, die dann mit hochmütiger Verachtung in ihren Kraftwagen an dir vorübersausen? Wir denken:

Nein!

Die Hotelerie nennt das Reiseauto oder richtiger gesagt das Luxusauto ihre Rettung. Auch wir bedauern, hier Stellung gegen die vermeintlichen Interessen der Hotelerie nehmen zu müssen. Aber das Unterland sagt uns, dass alle Liegenschaften und Gebäulichkeiten an den Autostrassen stark entwertet werden. Sollen alle die vielen kleinen Anwohner an unsern Strassen dieses Opfer bringen zugunsten der Hotelerie? Soll der Kleine leiden, damit der Grosse **vielleicht** wieder zu seiner Sache kommt? Willst du das, Bündnervolk? Wir hoffen:

Nein!

Ein kleiner Riss soll heute entstehen in dem Damm, der uns schützt. Aber das soll nur der Anfang sein. Heute sollen es ein paar Strassen sein, morgen sollen es alle sein. Man will uns teilen, um uns um so leichter zu besiegen. **Drum Bündnervolk, wach auf; keiner bleibe daheim; jeder stimme wuchtig**

Nein!

Das Komitee der Kreisversammlung V Dörfer.

Pidoll is one of the very few authors of the time who recognized the far-reaching social consequences of the automobile, with its increasing monopolization of streets and surfaces, to the exclusion of nonmotorized travel and the public sociability that went with it. Because the street looked like a public space—which, like the common in earlier times, should be accessible to everyone, whether for strolling, playing, window-shopping, or travel—the automobile threatened to decrease the territory devoted to life's public diversity. Should the "masters of time and space" be allowed to indulge their taste for speed, thereby curtailing the public right to the common use of the streets? Pidoll said no.

> The view that the definition and function of public ways and streets should be left entirely to traffic . . . is wrong and unjustified. In particular, the streets and squares of the metropolitan areas and towns are more than mere thoroughfares, like, for example, railroad tracks. They belong much more to the whole layout of the city; they are the population's site of settlement; they form the surroundings of the buildings, the milieu in which the personal, social, and economic life of the city in no small part takes place; and they offer satisfaction of the inescapable need of city dwellers to go for a walk outdoors. . . . Automobile traffic in its present-day form involves, as we have seen, the constant endangerment, disruption, and mobilization of passersby or other vehicles, as well as a severe infringement on those community relations that correspond to an advanced culture. It is legally and actually incompatible with the rights of common use of nonmotorists, the great majority of the population.[12]

The right to the street, the right of the public to remain unburdened and secure, was a cause for concern not only in Germany, but in car-crazed France as well, and in the United States, the model case of motorization. But nowhere do we find anything like a systematic weighing of the advantages and disadvantages of this new technology from a political angle. The power of established fact caused the protest to grow quieter over time and degenerate into grumbling, even if as late as 1911 such a respected periodical as *The Economist* still struggled with the troubling question "Is the loss of agreeableness and comfort, is the burdening of the public to be justified by the pleasure enjoyed by comparatively few individuals?"[13]

12. Ibid., 63f.
13. Quoted in William Plowden, *The Motor Car and Politics, 1896–1970* (London: Bodley Head, 1971), 95.

MUNICIPAL CHANCELLERY OPPOSES THE AUTOMOBILE

On July 12, 1923, the Building and Forestry Department in [the eastern Swiss commune of] Chur received a letter from the Municipal Chancellery of Domat/Ems containing the following:

> We find ourselves presently forced to lodge a complaint with your department owing to infringements of automobile traffic regulations.
> 1. On the 6th of this month, an automobile drove through here at 11:30 A.M. It was not possible to note the number.
> 2. On the 7th of this month at 12:20 P.M. we had the same pleasure, auto no. 2277A. It came from Splügen. We permit ourselves the question of whether the legislative bill, so carefully worked out, does not foresee arrangements to discourage such "stray" drivers. It is utterly inconceivable to us that an automobile can drive on a closed street from Splügen to Chur without being stopped. We demand that the owner of the automobile be found and made to answer. You will kindly keep us informed of the progress of the investigation. To us, and, we hope, to higher governmental instances, this is a matter of putting an immediate stop to infringements. We owe as much to our citizens.
> 3. The last straw, however, came yesterday when, between 2:00 and 3:00 P.M., an automobile coming from Chur drove through here, a secretary from the government offices sitting in same. Would you like to inform us, care of the people, as to the mission for which the gentleman in question was traveling by automobile? Did he perhaps wish to undertake a pleasure tour of the automobile-loving uplands?
> 4. A general complaint is also lodged herewith against the motorcycle from the telephone office, which drives through almost daily. To the best of our knowledge, motorcycles are prohibited in Graubünden.
>
> Respectfully yours,
> By the Municipal Council of Ems: A.F.

The Germ of Future Greatness

After years of debate over what to make of the automobile, this newest creature of technological progress, the time came for the political authorities to try to introduce order into the tangled relations. Things got moving at the national level when, from 1906 to 1909, the German parliament struggled to navigate an automobile liability bill through intricate legislative channels. In light of all the accidents and various annoyances, popular representatives felt called to action: How could the German people be spared the obvious harm? Was it not right and proper to obligate automobile drivers to take responsibility for all potential damages? Demands were set and deals were made, the battle of words producing both populists and lobbyists. The divergent interests met head-on in the legislative chambers: angry conservatives, battling on the side of the people, insisted that drivers be held liable in principle, unless they could prove otherwise, while automobile associations, fearing financial strangulation, pressed to avoid legal regulation altogether. Only the Social Democrats appeared to have no opinion on the automobile; their concern in the matter went first of all to regulating the work hours of chauffeurs.

In the balance of pro and con an argument soon surfaced, ever more insistently, which those concerned with conditions on the street could scarcely counter. While they continued to lodge complaints about overturned wagons and disturbances of rural peace, their opponents changed their tune, taking up the national anthem: the well-being of German industry was in question, and who could remain indifferent to that! This argument suddenly moved the discussion into another dimension, one that had nothing to do with the advantages or disadvantages of driving a car. Eager to accommodate, the *Allgemeine Automobil-Zeitung* opened its pages to the pleas of the machine manufacturer Nacke from Sachsen:

> I took up automobile manufacture primarily for two reasons. First was my purely technical interest. Second, I wanted to do my part to contribute to the creation of a German automobile industry, to open a source of worthwhile earnings for our German machinists, and to help our fatherland and our national commerce retain those moneys which are now being lost abroad in the purchase of French automobiles. . . . The heavens themselves would register the injustice done an industry struggling painfully to survive, were it to be crippled by a high tax on automobiles just because the nation needs money and the parliament, always hostile to the automobile, will probably approve such a tax without hesitation. Half-tamed countries give new industrial ventures freedom from taxation for years to come, but in highly civilized Germany there is so little concern for the situation of a new industry that it is squeezed to death through taxes, which affect sales, even before it has the chance to develop.[14]

Whether in the question of taxes or of liability, attempts to counter the social damage caused by the automobile ran up against an image of society in which only one characteristic of the automobile truly counted: it was a profit-making commodity with markets that German industry could ill afford to abandon to other nations. Only the domestic manufacture of automobiles would stave off decline and keep German money at home, money that, by nourishing both consumption

14. *Allgemeine Automobil-Zeitung*, 1908, no. 3, 33f.

Cover of The Motorcar, *Magazine of the Central European Motorcar Association, 1899.*

and investment, would ultimately lead to the expansion of productive capacity. In the preceding few decades the nation's greatness had seemed to grow along with its industrial base: great power ambitions were no longer, as in the preceding epochs, founded on "the divine right of kings," "rational state order," or "diligence in commerce"; now what mattered was dominance in the output of mass production goods.

In the course of the nineteenth century, the "normative" social order of old Europe had given way to a "functional" social order, one that privileged values, classes, and forms of production according to their contribution to the output of goods. Throughout Europe factory installations and combines arose, while the first technical analysis of the problem of growth appeared in 1904. Banks got involved in investment capital, and the epoch became aware of its specific origin in the "industrial revolution." Were not those who stood in the way of the budding automobile industry mere grumblers? Who could continue to

> But who are those people who cry to the state for help, who appeal to
> the legislators against motorists? They are the same people who half a
> century ago wanted no gas lighting, who petitioned the Prussian king to
> prevent the construction of a railroad from Berlin to Potsdam. There
> will always be people of this sort, people who refuse to see that the
> better is the enemy of the good. To these same people today, the auto-
> mobile is the very personification of the idea of progress, and because
> they battle the latter wherever it appears, they must also campaign
> against the gasoline-driven monster. These people are now crying out
> for the state to use its cudgel, as on other occasions they prefer to cry
> out for the safety and comfort of the government crib.
>
> *Allgemeine Automobil-Zeitung,* 1906, no. 16.

dwell on damages and waste, on reckless speed and disturbances of the
peace, when the greater good of the nation hung in the balance? "The
infant automobile industry," the *Frankfurter Zeitung* determined,

> cannot today endure any experimentation at all, not to mention any
> tests of strength. Precisely because it bears within itself the germ of
> future greatness, it should be sheltered and defended by the state, since
> it will doubtless be called on to play a significant role in the economic
> life of our people. . . . It does not yet possess that settled foundation
> necessary for its emergence in the world as a successful competitor. It
> must first sink roots in its own native land; its native land must offer it
> the conditions of profitability, so that it can expand from its secure
> center at home into other markets.[15]

The goal of greatness demanded that the nation place itself at the
forefront of progress, the costs of which to daily life, as the industrialists understood the issue, could be neglected in favor of the promises of the future. The Wilhelminian era—contrary to its reputation—
was eager to modernize, and indeed, one new technology after another
kept springing up. Little technological miracles became the topic of the
day, and daring engineers its model figures. It was the classical age of
progress. Whoever had the money could acquire electrical light in the
1890s, enjoy a new mobility with the bicycle, write letters on a typewriter, and admire the first moving picture shows. A few years later it
became possible to undertake a Freudian analysis, fly in an airplane,
and study the principles of a jet engine or even the fundamentals of
space travel. The twentieth century began quite literally at the turn of
the century.

15. Quoted in ibid., no. 10, 41.

The upwardly mobile middle class, the engineers and white-collar workers, regarded themselves in particular as the carriers of history. Progress, in whose service they defined their mission, unfolded in their eyes naturally and purposefully, like a flower. And history, more precisely the most recent history, appeared to them as an upward movement, its direction indicated by the heavenward-pointing Eiffel Tower: higher, faster, farther. The automobile presented itself as "the personification of progress altogether"; thus it was clearly right that it prevail over the objections of the "retrograde" and "philistines." Those who failed to understand would soon enough be forced by the power of established fact to see the error of their ways, as Baundry de Saunier had already stressed at the turn of the century:

> The mechanical carriage is here once and for all, and, however one might persecute it, it will not die, because it corresponds to the logic of economic progress and the needs of our time. To resist it is precisely as foolish as to struggle against time, age, the work of the human mind, perpetual motion, against the forces of nature, which exercise an influence on us and which our bad mood can never alter.[16]

Happiness—it will ultimately be achieved by the march of progress; whoever brakes material progress runs the risk of hypothetical retrogression. Understood in this way, the industrial apparatus of the nation acquires a historical dignity. It secures for the nation, against the lurking opposition of other countries, a favored place under the sun of the future.

What the critics of the automobile saw themselves confronted with in the debates of the time could be called the executive syllogism of competition-driven progress: (a) technological development cannot be stopped; (b) escape is not an option, so Germany must take the lead; (c) therefore, we are called upon to support the automobile and its industry with all the means at the state's disposal. In the face of so strong a consciousness of national responsibility, the critics became subdued and saw their questions—whether the automobile was necessary at all, and whether its advantages outweighed its disadvantages—grow strangely insignificant. Discussions of the practical value of the new product fell by the wayside. The world market cast its long shadow over debates about the meaning of motorization on native streets.

16. Saunier, *Grundbegriffe*, 14.

Whatever the nuisances and dangers they might point to, given the commandment of industrial might, the prudent and cautious could not but appear to be enemies of the nation and enemies of progress. Nor were the objective-necessity strategists lacking in answers to their concerns: since the automobile was indeed here to stay, motorists, pedestrians, and wagon drivers—so ran the demand in many speeches—had to be trained in proper behavior on the street. At the Extraordinary Automobile Convention in 1908 in Berlin, the keynote speaker explained why the only viable perspective was that of the motorists:

> Improvement cannot be achieved by a liability law, but solely through public self-education, through a general awareness that the motorcar is not a useless, aggressive sports vehicle, but already today represents a recognized and extremely important means of transport for persons and cargo that will never disappear. . . . The fright and consequent mindless behavior of passersby sometimes leads to argument. Most people lose control of themselves when they hear the warning signal and instead of standing still begin walking all the more rapidly. . . . The public is gradually becoming used to looking both left and right before they step into the roadway. To be sure, accidents happen there too. It is, however, strange that people always want to shift responsibility for the frights they take and the accidents they suffer onto other people and that they abuse the automobile driver, who just happens to be there. They would do better to take themselves in hand and say to themselves, You've done something stupid, you've gone and crossed the street without looking both ways.[17]

It is not, the speaker concluded elsewhere,[18] as if people can behave on the street as if they are strolling in the parlor or retiring to the bedroom!

With the automobile's inevitability taken for granted, there began to develop what might be called a traffic education discourse. Drivers have to be "qualified"; pedestrians should behave "correctly" and not "wrongly"; carriage drivers and cyclists have to show "consideration"; and one and all, in view of the new requirement for discipline, are to be seen as in need of training. Only in this way could an "order" be created on the public streets that would minimize the dangers of motorcars. "The majority of accidents occur precisely because other types of street traffic are altogether unwilling to accept the new conditions brought about by the introduction of the motorcar and to conform to them." "Unbelievable," an outraged doctor pronounced in an

17. *Allgemeine Automobil-Zeitung,* 1908, no. 11, 39.
18. Ibid., no. 10, 38.

What is actually happening can best be determined from, in addition to personal observation, the catastrophes involving the automobile and its associated circumstances, as well as from the relevant court proceedings.

The ghastly chronicle of these accidents is inexhaustible. *In Vienna alone in the first half of 1912, 438 automobile accidents* have occurred, in which *sixteen people died, and from the end of May to the end of September 1912—that is, in four months—seven children have been killed* by automobiles on the streets of Vienna. *Does the blood of these victims not smell to high heaven?* Numerous automobile accidents, as a result of the well-known arts of concealment, do not become general knowledge at all or, if they do, are not appropriately represented. What is characteristic of them is that the force and speed of the automobile are constantly causing especially severe injuries. The following account cites, in the briefest terms conceivable, a number of automobile catastrophes that have occurred in Austria and abroad, respectively sixty and twenty, with few exceptions in the years 1911 and 1912 alone.

One dead, two severely injured. On Whitsunday in the afternoon, an automobile being driven by the owner swerved on a curve into the roadside ditch, after it had passed by the market square in Öd "at a high rate of speed," and collided with such force against two pear trees that the driver was thrown backward, fell under the rearing automobile, and lay there severely injured. Of the other passengers in the automobile, one suffered serious internal injuries (hemorrhages of the lungs) and another emerged with a broken arm. The automobile had been traveling for stretches at a speed of 100 kilometers per hour, and traversed the 125 kilometers from Vienna to Amstetten in one and three-quarters hours. It is to be noted herewith that the Orient Express from Vienna to Amstetten requires two hours and two minutes, that is, seventeen minutes longer. (*Morgenblatt,* June 1, 1909.)

Two dead, two with life-threatening injuries. A driver with three waitresses and one mechanic in the car was on the Franz Karl-Straße on the outskirts of Vienna. The automobile suddenly collided with the street barrier. The planks on the barrier, thicker than a man's arm, were broken in two places, the ground deeply dug up, and the automobile demolished.

One dead. An automobile was on a test drive on the Schönbrunner Hofallee in Neu-Erlaa. It collided with a tree and was demolished. A passenger was killed. (*Morgenblatt,* April 19, 1912.) The speeds attained in "test drives" are well known—and this on public streets.

Three severely injured. A student was driving (on December 16, 1910) from Vienna to Semmering Pass with two ladies and a gentleman in the car. On the road to Neukirchen he attained a speed of sixty kilometers per hour. The automobile began to spin and suddenly, just after it passed the second contractor's wagon standing in the street, was cast along a giant arc, so that it went off onto the side-path. There the back wheel dug deeply into the ground, ripping it up. Nearly simultaneously, the automobile flipped over *twice along its lengthwise axis.* The damaged wagon came to a stop on the side of the street, leaning up against a tree. The passengers in the automobile, who had been thrown out, lay injured beside the street.

One of the ladies lay unconscious for weeks and in April 1912 still had to be pushed in a wheelchair because her foot was completely crushed. The second lady was less seriously injured, but was capable of returning to work only after a year had passed. The male passenger (a chauffeur by trade) suffered severe injury and was incapacitated for a year; in April 1912 he was still walking on crutches. At the ensuing trial, the driver declared himself not guilty. He claimed he was not driving fast, but that the road was very muddy and that the automobile went into a spin because of the street conditions. The spin, he continued, was such that the automobile was thrown against a tree, whereupon it flipped, thus causing the accident. According to the testimony of the passengers, however, the automobile was traveling at a speed of sixty kilometers per hour as far as Traiskirchen, and later at as much as seventy kilometers per hour! Before the accident occurred, the automobile skidded twice onto the wrong side of the road. (*Morgenblatt,* April 26 and July 4, 1912; trial transcript.)

One dead. A twenty-six-year-old man was knocked down by an automobile on the Rennweg in Vienna and run over. He suffered numerous fractures of the skull and died a few minutes later. (*Morgenblatt,* May 1, 1912.)

At the ensuing trial (cf. *Morgenblatt,* July 27, 1912), the driver was sentenced to two months' close confinement, with the reasoning that he was driving very fast and had given no signal. Traffic was heavy at the location in question, and there is no doubt that he failed to exercise the necessary caution.

Two dead. On Whitmonday in 1912, an automobile was proceeding along Praterstraße in Vienna. "The speed was very fast." A woman accompanied by a young girl wanted to cross the street in front of the Karl Theater. She could not get out of the way and was caught by the automobile and thrown to the ground. The automobile drove over her body and then stopped. Agitated cries were raised against the driver. There was a large pool of blood on the street. The women had a fracture of the base of the skull and broken ribs on the left side; the girl suffered internal injuries with bleeding. The woman died on the way to the Rudolf Hospital; the girl passed away shortly after being admitted. A witness to the accident saw how the two of them pulled each other back and forth, obviously confused and undecided as to where they should go to save themselves. There was very heavy traffic, especially automobile traffic, in both directions. The girl suddenly pulled her companion toward the center of the street, directly in front of an automobile neither of them had seen. According to the witness's testimony, he has already witnessed two train accidents; but the sight was not as horrid, as terrible and arousing of compassion, as this one of the automobile accident. The driver was arrested at the site of the accident. (*Abendblatt,* May 28, 1912.)

Is one to ascribe the fault to the two poor victims, thereby in a way legislating that the death penalty is to be exacted for fear and helplessness on the street? Is it not much more the case that when express vehicles are allowed to storm into *a chaos of people and wagons,* as is certainly to be foreseen on such a busy street as *Praterstraße* during *Whitsuntide,* the probability of an automobile catastrophe rises to a certainty?

> Michael Freiherr von Pidoll, *Der heutige*
> *Automobilismus. Ein Protest und Weckruf*
> (Automobilism today: A call to protest) (Vienna, 1912).

essay, "unbelievable is the carelessness with which the public continues to cross the busiest streets and the way many parents allow their children to use the streets as playgrounds, as if no such thing as the streetcar or the automobile existed." [19] Before it was even possible to think of transforming the space devoted to streets into thoroughfares, a transformation of behavior was essential: training in attention and self-control had to be second nature for everyone, and a protective armor of foresight and automatic reflex was a matter of necessity.

> It must become habitually ingrained in the entire population to use the roads as little as possible wherever sidewalks are available, to look left and right upon entering the road, to walk and drive only on the right, even when the whole street is empty, and not to stand around in the street. A certain amount of traffic education for the population is urgently needed. [20]

Respect for the automobile had to be introduced into daily perception, into the core of habit—just another little bit of civilization, which, once gained, would allow progress to make its home in Germany and bestow upon the national economy a bright future.

19. Ibid., 1909, no. 11, 27.
20. Ibid., 1906, no. 5, 74.

Cars for the Few —
Cars for the Many?
(1920–1933)

The young ladies of the bourgeoisie danced the Charleston, skirts were short, and rhythm was the name of the new freedom of the dance halls on Berlin's Kurfürstendamm. Meanwhile, the wives of workers were standing for hours, lines in front of the stores snaked along the street, and mouths in the working-class suburb of Wedding were hungry. Life in the twenties, the golden years, presupposed the poverty of the majority. There were opposite worlds in automobiles too: Bugatti and Citroën, Cadillac and Model T, Mercedes and Dixi, limousines for those at the top and tiny cars for those down the ladder. Driving—would it remain an exclusive form of play, or could the masses be motorized too?

The Beauty of Function

After 1924, as inflation was overcome and things picked up again for a few years, new strata of potential buyers were drawn to the automobile. Particular groups—the newly arrived, like management personnel, engineers, or doctors and lawyers—placed a high value on acquiring a car, the symbol of modernity. They could at least compensate their deficient sense of status with the consciousness of progress, thereby underlining their claim to a position at the top. Even those well-heeled sorts who had not trusted motorized devilry at all before the war relinquished their reservations and joined the trend of modern times. The 1924 count of 130,346 automobiles had climbed to 489,270 by 1932: about 1 percent of the German population now had access to a car. Although the motorcar by no means commanded a mass market, nonetheless it was outgrowing its role as a sports item for technology enthusiasts and becoming a consumer good for seekers of status. Design changes provide a legible record of this developmental trend. It became imperative that manufacturers apply their ingenuity to the operational convenience of their vehicles, because the number of adventurers ready to welcome the challenge of a flat tire or a torn brake cable was growing ever smaller. "More consideration for the gentleman driver," the first issue of *Der deutsche Automobilbesitzer* (The German automobile owner) demanded, for

> traffic increases every month in Germany, and this increase is matched by the growing number of gentleman drivers who take over the wheel of their own car and sometimes even the maintenance. . . . For it is a common experience that once one sits behind the wheel, one no longer likes to ride in the back of the car. . . . The automobile is justly becoming available as well to broader social strata, so that today such people as are not blessed with an abundance of luxury goods can afford a car. . . . Will economics, trade, and technology do right by the gentleman driver? The automobile industry made the first move toward the gentleman driver: it has created a type of car that is easy to steer, elegant in appearance, and undemanding in maintenance. One cannot expect of a gentleman driver that he know as much about his car as the professional driver who is also a trained mechanic.[1]

If a technological product is to become a mass commodity, it must forsake its identity as a specialist's toy and become, to employ the sales jargon of today, a user-friendly instrument. Its functioning must be made relatively free of surprises by letting the technical design handle tasks for which the operator was formerly responsible. Accordingly, in automobiles of the twenties battery power made the magneto-ignition obsolete, a four-wheel hydraulic system replaced the cable brakes, and electric lights took over from carbide lamps. Windshield wipers and rearview mirrors also appeared, easing the task of driving. Furthermore, independent wheel suspension put an end to the steering wheel shudder caused by fixed axles, just as "floating" engine mounts contributed to increased driving comfort. And while the open touring car

1. *Der deutsche Automobilbesitzer* (The German automobile owner), 1928, no. 1, 3.

"Brennabor: The most beautiful German car." Advertisment for the International Automobile Show, Berlin 1928, from Die Dame, *November 1928.*

made of a sheet metal jacket on a wooden frame was still very much in evidence at the beginning of the decade, it was the luxury wagon with an enclosed passenger space stabilized by a steel frame that set the standard by the end.

This "squared-off" styling, with the horizontal lines of the long hood, the waist-high rim of the body, and the high roof set off against the eye-catching vertical lines of the radiator and the overall boxlike construction, practically became the symbol of the time. Anyone who saw an Audi Zwickau rolling along the street was unavoidably impressed by its balanced proportions and closed cubic form. Such a car was representative not only of its owner but also of the very spirit of modernity that informed the vision of the epoch's technical-artistic elite: to seek beauty in a thoroughgoing objectivity. No longer were embellishments added or decorations pasted on; instead the functional form itself was considered a complete expression, so much so that, as Le Corbusier suggested, a fine automobile could assume a rank equal to that of the Parthenon.

Up until now one criterion of beauty had been "uselessness." From now on a lamp or a locomotive could be perceived as beautiful, as long as its form expressed unadorned function: the more strictly functional the machine, the more beautiful its formal aesthetic. The automobile of the twenties reflected in its flat surfaces and geometric shapes the social project that the elite cadres of form, from Muthesius to Gropius, had in mind: to express the purity of the technical norm as the simultaneous guarantee of beauty and rationality. By replacing the decorative carvings on chairs with a steel pipe construction, transforming the unorganized large kitchen into a small one that conformed to routine movements, or attempting to balance the chaos of cities through a spacious separation of work and residential areas, they hoped to introduce the fresh breeze of reform into society. Objectivity was the order of the day, a bold front against the philistines and reactionaries stuck in the stifling atmosphere of the imperial period and still dreaming of twisted mustaches and pointed helmets. And objectivity headed the list of ways to show the class strugglers on both sides of the barricades how national divisions could be reconciled through carefully planned order.

In following this trend, automobile design proclaimed the hope of the time, namely to bring the heavenly technical civilization of the thinkers down to earth and let reason finally find a home in stuffy Germany. This message invested in the automobile clearly resonated with the bourgeois strata, who, as the flag bearers for the modern cause, were only too pleased to find a solid foundation for their hegemony after the collapse of the monarchy. They wanted to transform society in the spirit of technology; the automobile embodied this project. They claimed priority over officers and the agrarian aristocracy; the automobile underlined their superiority. It encapsulated their self-image. The possession of an automobile made palpable what one stood for, the aim of one's ambitions—and the measure of one's status.

The World of Ladies and the Spirit of Extravagance

The automobile sank its roots into society from the top down. The upper classes were of course the first to enjoy the delights of vehicle ownership; indeed, not until 1970 were more than half of German working-class families able to taste the pleasure the upper classes had been parading before them for fifty years. Correspondingly, the automobile fleet did not grow from the motorcycle through popular cars to the status vehicle, but vice versa: at the beginning was the luxury automobile, to which all kinds of middle-class and compact cars were added over time. Thus it was that the luxury market, rather than mass consumption, was the midwife of the automobile, for into the 1930s the industry geared its production above all to customers who looked not to price, but first of all to beauty and performance. Scarcely anyone conceived of acquiring a private car because it might be useful or even necessary. On the contrary, one purchased an automobile to cultivate a pleasurable lifestyle beyond the daily routine. A Hispano-Suiza emanated elegance and power; it was in demand as a means of gratification, not as a means of transport. And even after an occasional inexpensive "Laubfrosch" (Tree Frog) could be seen clattering around the corner, it was still the large automobiles with their wealthy owners that put the sheen on driving and created a model for the masses.

But what was it that inclined these gentlemen, beyond all considerations of utility, to spend their money on such a luxury? How did driving become a matter of elevated taste? A 1926 article entitled "What Car Do You Drive?" in the magazine *Elegante Welt* (Elegant world) provides a first indication:

> Who fails to remember the recurring questions (favored especially among the ladies) of that time when the woman ruled "house and hearth" and all conversation revolved about the servants? That is when turns of fashion began to be the characteristic feature of life. One spoke only of seamstresses and the dressing table, or possibly of bobbed hair. . . . Only now does a new concern join the list. When two ladies meet they naturally survey the aforementioned subjects down to the tiniest detail—but it does not stop there. On their morning walk into town, for shopping or even just to work off a pound or two, the conversation babbles pleasantly along. Suddenly a charmingly

Arriving in the world of cultivated pleasure. Poster, 1924.

enclosed two-seater glides by. The vehicle breezes away like a breath of air, a small hat in Bordeaux red projecting just above the steering wheel, and the eye is left with nothing but the image of a big spare tire encased in shiny leather, which remains just a moment longer to mark the disappearing car as utterly elegant.[2]

The automobile had become material for conversation in better circles. Now that their tastes in clothing fashions, dining style, and parlor furnishings were fully refined, a new product, the automobile, was incorporated into luxury consumption, into the culture of gratifica-

2. *Elegante Welt* (Elegant world), 1926, no. 21, 25f.

tion, which itself was vitalized, above all, by the society of ladies. "In the time of chauffeurs and open wagons," a French author explained,

> women were not in love with the automobile, or at least, it was not a true love. . . . Today this is not true in the slightest. Since the fairer sex has been granted the comfort of easy entry at the side, upholstery, suspension, and protection from the weather, without threat to hair and makeup, since then the women love the automobile. . . . For them the automobile is something that shines, costs much money, and therefore is very chic. It affords them a new luxury, a new-fashioned means of inspiring a bit of envy in their girlfriends who can manage nothing more than the subway or a carriage. For the manufacturers it is a decisive turn that they, the women, induce the masters of creation to purchase a large number of automobiles.[3]

The association between ladies and the automobile first cleared the way for the establishment of driving as a model of consumption. At no other time does one find so many placards and commercials that present women (better, ladies) together with cars. The automobile was slowly invested with the emotional aura of consumption, for it was the figure of the lady that embodied the domain of private gratification.

In the course of the nineteenth century the sphere of the woman had arisen to oppose that of the man. The division between home and the world of professional work came to be reflected in a polarization in the world of sentiment as well. The man, at least to hear him talk, was concerned with a sense of duty, accomplishment, and thrift; but the (bourgeois) woman developed in opposition a consciousness of taste, leisure, and lifestyle. The bourgeois parlor, with its wallpaper and carpets, its sofas and curtains, was testimony of the dichotomy, as were the trimming and frills, the velvet and satin in clothing fashions. Consumption occurred primarily in the private realm, in which the woman set the standards.

The rising middle class drew its nourishment ultimately from two underlying cultural trends: the Protestant ethic, to use Max Weber's term, which furthered production through thrift and a drive to perform; and the spirit of extravagance, which, through sensual satisfaction and vanity, stimulated consumption. Not only the entrepreneurial spirit and factory discipline but also extravagance and a joy in buying had to triumph if true consumers were to be made of self-sufficient citizens. In this process the values of consumption crystallized around the

---

3. L'année automobile, *L'automobile, cette folie!* (The automobile, that madness!) (Lausanne: Edita Vilo, 1982), 112.

Lady with automobile. Cover for Vogue *by G. Lepape,*
1930.

nineteenth-century lady; it was not by accident that Emile Zola, in one
of his novels, named the first department store in Paris "Ladies' Paradise." Or, to quote the sarcasm of Schopenhauer: "Females think in
their hearts that the purpose of the male is to earn money, and theirs,
on the contrary, to squander it."

The world of ladies embodied the values of consumption, while the
world of youth embodied the new. Thus in early advertising illustra-
tions it was young women who summoned customers to purchase new
consumer goods, whether a cigarette, a bicycle, or, indeed, an automobile. The industry came along as the fulfiller of feminine values, so
that family life—the sphere of consumption—could be rendered economic. This is why the automobile, with the help of the ladies, appeared to the astonished masses of the twenties as a shining article of
consumption.

The automobile indeed conquered the people's imagination, but it far outstripped the realities of the purse. It had taken twenty-five years for the car to become, at least in fantasy, a commodity for mass consumption: from "sporting pleasure of the rich" to "cars for everyone." In the 1920s the automobile won a permanent place in desire; its mechanical beauty seemed to be a vision of the future. It was still true that the man on the street got only dust and noise from the automobile, but for many the fascination it inspired began to outweigh the irritation. The grapes no longer hung too high for the spectators; they wanted to become actors themselves in the drama. Why should something that was the right of the rich not be cheap enough for everyone? With their demonstrative possession the wealthy were proving that money and no longer birth formed the foundation of social prestige. The time was past when life's offerings were already decided in the cradle. Had not everyone the right to do as the elite did?

People cast their glance across the great pond to America, where yearly production in 1923 already exceeded three million vehicles, allowing visions of the future to arise that burst the bounds of the established image.

> Things will develop with the automobile as they did for the horse, the railway, and the bicycle. Not the grand automobile, which for a long time, if not forever, will belong only to a small, privileged minority, but the middle-sized and especially the small car. . . . The day will come—more quickly than we think—when everyone will have a place beneath the house in the garage (itself a matter of course in the future); when the automobile—who knows?—will be included in the price of the rent like other modern comforts; when everyone keeps an automobile (that means speed) at home, just like our houses now—to the bewilderment of our fathers, could they but see it—put water, gas, electricity, bathrooms, and central heating at our fingertips.[4]

This text signals a new relationship to the automobile: no longer an awe-inspiring luxury good, but an article belonging to the basic equipment of the modern person. Although there were only 130,000 vehicles on the streets of Germany, an image of full motorization was forming in which everyone would park easy-to-use speed in the garage. The new society conceived of itself democratically, so that for farsighted minds motorization meant *mass* motorization. Those more

4. *Automobil-Revue*, 1923, no. 7, 126.

The manufacturers are demanding an unheard-of increase in the tariff on foreign cars. This added expense will naturally be borne by the German automobile buyer—the chance for countless Germans to acquire a car will be dashed. . . . The vitalization of the automotive industry is more important to the workers than to the manufacturers. For the more motor vehicles there are in service, the more employment opportunities there will be for qualified chauffeurs and repairmen and for unskilled workers like garagemen, street builders, and such. For every automobile produced by a genuine motor vehicle manufacturer, three or four people gain positions and their daily bread. Consequently, an increase in the number of motor vehicles is a clear plus for the labor force and for the economy as a whole. This so highly desirable plus will be achieved more quickly the less expensive cars are, because they then penetrate into the lower income brackets.

This circumstance makes the labor force less concerned with where the cars come from; the main thing for the workers is that many cheap cars come in the first place. . . .

It will not be otherwise in Europe. Here, too, Ford will get started with higher wages—or he will not get started at all. With the rise in purchasing power for the masses, everything else follows as if of itself. Once automobile drivers have become a mass, even the pettifoggery of the legislators will begin to decline.

The motor vehicle, that fun-filled, purring, and devilishly fast factory on wheels, has changed our entire public and social life in a few short decades; it has made people more mobile, distances shorter, usable land more extensive, besides having wrought a powerful transformation on the technical organization of work methods. Now the transformation will continue in Europe, notably in the area of wage policy. And this will, in turn, foster an increase in the income of the masses and their social advance. . . . This newest change will not cause harm either. On the contrary, the revolutionary automobile will serve the cause of the revolutionary working class.

Metallarbeiter (Metalworker), no. 10 (1930).

conservatively inclined, in contrast, sensed "equality mongering" and "leveling" in this inherent technological tendency, because they saw the status-conscious way of life, vouchsafed to only to a few, being thrown overboard.

To be sure, the economic foundation of this utopia of motorized equality was still lacking: neither were cars cheap and mass-producible, nor did the public, however eager to buy, have sufficient cash in their pockets. But the solution to this dilemma was already visible across the Atlantic: Henry Ford, the new idol, appeared to be breathing real life into the American dream. In Detroit in 1913 and 1914 he introduced the assembly line, along with the eight-hour, five-dollar workday, thus increasing output through serial production and, with the higher wages, raising purchasing power. It is therefore no surprise that Ford's autobiography became a bestseller in Berlin in 1925, and that it opened up new horizons. For Ford recognized that workers were also consumers, who, if given higher wages, would move into the position of buying the automobiles produced on the assembly line. A new social project was tied up with industrial mass production: the consumer society. Practically no one in Europe saw this so clearly as André Citroën, who acted accordingly by following the lead of Ford's Model T and producing his small "Citron" in great numbers: "Fast, good, and economical production becomes a necessity if one wants to reduce the price and bring the goods produced within reach of the largest possible number of consumers."[5]

In Germany, such role models could be viewed only with envious eyes. Industry had been weakened by war and fragmented into many firms; customers, short of money, could not pay even the considerable motor vehicle taxes and gasoline prices; and domestic sales got into real trouble when lowered tariffs after 1926 allowed foreign cars to flood the market. During the decade a few attempts were made to offer the German public a small car, a "proper" automobile at a manageable price. All kinds of curious minicars came on the market; none, however, blessed its creator with long-term success. In 1924 Opel introduced assembly-line production and copied, almost centimeter for centimeter, the "Citron"—without license, but with the green paint job that gave it the name "Tree Frog." People were amused at the little Hanomag—"Two pounds of tin and a pound of paint and the Hanomag is done"—calling it "army bread" because the sheet metal was rounded in the front and rear and even projected over the wheels. And the BMW Dixi, which began production in 1928, swayed more than it drove because of its narrow chassis and high body, and with a top speed of 60 kilometers per hour could scarcely induce a motorcyclist to make the switch to a car.

The problem of the small car had not been solved, in terms of either design or economy, when the international depression set in. Thus L. Betz, in 1931, had simply to recall the unrealized program in his polemic against the German automobile industry:

5. Quoted in Sylvie Van de Casteele-Schweitzer, "André Citroën: L'aventurier de l'industrie" (André Citroën: The adventurer of industry), *L'histoire*, no. 56 (1983): 13.

"A day's production at Opel." Advertisement from Elegante Welt, *1925, no. 13.*

A people's automobile will be developed on the basis of neither motorcycle nor luxury car construction, but only on its own terms. We have here a vehicle that must be free from tradition and compromise, that can have only one model: the Ford. Not the Ford in its present-day design, but as an idea. The miniaturized passenger car is just as inappropriate as the enlarged motorcycle as a basis for design. The one will

"I love only one . . . ! Hanomag, the little one." Advertisement *from* Berliner Illustrirte Zeitung, *1926, no. 37.*

be too heavy and too expensive to produce, the other ill suited in traffic and unusable for its purpose. . . . No hopping tree frog, but a car designed for the street, offering a maximum of comfort but a minimum of luxury.[6]

The Vision of Penetrable Space

Mass motorization was on the agenda at the beginning of the 1930s in Germany, but not only was a people's automobile lacking, the streets

6. L. Betz, *Das Volksauto. Rettung oder Untergang der deutschen Automobil-*

were hardly in a condition to tolerate a nation on wheels. This spatial aspect of motorization had crept into social consciousness already in the twenties; the driver, after all, required a passable network of streets just as a fish needs water. Yet despite some tentative attempts at an infrastructure—the first gasoline pump in Hamburg in 1923, the first lighted signal lamps in Berlin in 1928, and in 1929 the first parking garage—the condition of rural roads could only be deplored: they were narrow, twisting, dusty, and, naturally, too few. Such streets could easily take the fun out of driving, particularly for the owners of faster cars. The old rural roads were grounded on a different conception of space: they were connecting lines between neighboring locales, not thoroughfares to distant destinations. Planned according to small scales, laid out for slow speeds, twisting along creeks and over hills, and emptying directly onto a market square, these streets were of use to bicyclists and horse carts but not fit for the space-mastering power of the automobile. The automobile operates according to a different standard of distance; it comes into its own only when space is penetrable and speed unimpaired by curves, carts, or rubbish heaps.

It began to seem that cities and countryside should yield to roads built for speed; now, in the twenties, the first images of a spatial order defined by traffic start to appear. H. Kluge, in his inaugural speech as rector of the University of Karlsruhe in 1928, demanded the right of way for automobiles in the German states,

> where no one can claim . . . to have satisfactorily disposed of the traffic problem, and where not only an unfavorable influence on speed but also the further proliferation of motor vehicles is becoming evident. But when such is the manifest need, then right of way in the literal sense must be introduced, with neither cities nor rural areas falling behind in development relative to others. The large cities, and even the middle-sized cities, will sooner or later have to undertake street reconstruction, open up thoroughfares, build over- and underpasses. And even these [improvements]—since the cities as they now exist can be fitted to the new means of conveyance only over the course of decades—will not offer an adequate solution. Rather, new centers of traffic communication will have to be built, not merely owing to the increase in population, but also because of the inadequate traffic capacities in the old centers.[7]

industrie (The people's car: Salvation or decline of the German automobile industry) (Stuttgart, 1931), 45, 73.
7. H. Kluge, *Kraftwagen und Kraftverkehr. Kritische Bemerkungen über die bisherige und zukünftige Entwicklung des Kraftwagens und des Kraftverkehrs* (Power ve-

Professors in gown and mortarboard were no longer dozing through contemporary flights of philosophical fancy, as was often the case at official university addresses; large-scale proposals of a spatial organization suited to the automobile had become appropriate topics for festive occasions. The rector at Karlsruhe announced a project that lately occupied the minds of the technical elite: to straighten out the old-fashioned twists and turns of society and cover the country's surface with a network of streets conceived from a national point of view.

Le Corbusier achieved the appropriate perspective as he looked down from an airplane on the chaos of streets in São Paulo and found himself in possession of a new vision of order. As he put it in a lecture of 1930:

> We determined what a considerable amount of time is required to get from one point to another: little valleys, winding curves, hills, etc. On the ground we could appreciate the general topography, with its continual ups and downs, humps and hollows, and grew breathless in the tangle of streets seeking vainly to run in straight lines. . . . What if we construct a horizontal connection from hill to hill, from summit to summit, with a second connection to service the other major points? These horizontal connectors, running at right angles to each other, will be the great approaches and thoroughfares of the city. You will not fly over the city in your cars, but "roll over" it. These freeways that I propose to you are gigantic viaducts.[8]

The people's car and the freeway—the two pillars of a society on wheels—had already been conceived of at the beginning of the 1930s, and the ideas pressed for realization. They were, in fact, soon to arrive, but not under the sign of democracy.

---

hicles and traffic: Critical remarks on the past and future development of power vehicles and traffic) (Karlsruhe, 1928), 30.

8. Le Corbusier, *Précisions sur un état présent de l'architecture et de l'urbanisme* (Details on a present state of architecture and town planning) (Paris: Editions Vincent, 1960), 240, 242.

The Motorized
Volksgemeinschaft
(1933–1945)

Dictatorships live not only by force but also by emotional appeal; the shining eyes of the man on the street are as much a part of the image of the time as the Gestapo at dawn. A history of this enthusiasm in the period of German fascism has yet to be written. Yet whoever undertakes to eavesdrop at the corner bar and uncover that consent from below amid the oppression from above will have to make room for a chapter about the National Socialists' motorization policy. Adolf Hitler was the first German chancellor to honor the annual International Automobile Show in Berlin with a visit. Nor did he greet the visitors at the opening on February 11, 1933—less than two weeks after his appointment—empty-handed: "The motor vehicle has become, next to the airplane, one of humanity's most ingenious means of transportation," Hitler declared.

> The German nation can be proud in its knowledge that it has taken a major part in the design and development of this great instrument. If I have the honor today, on behalf of the president of Germany, to address the gentlemen gathered here and the industry, I do not wish to neglect the chance of communicating my conception of what must in future be done for this, perhaps the most important, industry: (1) removal of state representation of motor vehicle traffic management interests from the traditional management framework; (2) gradual implementation of tax relief; (3) commencement and execution of a

large-scale plan of street construction; and (4) support for motor sport events. Just as horse-drawn vehicles once had to create paths and the railroad had to build rail lines, so must motorized transportation be granted the streets it needs. If in earlier times one attempted to measure people's relative standard of living according to kilometers of railway track, in the future one will have to plot the kilometers of streets suited to motor traffic.[1]

At issue here was a program, not niceties; it was not by accident that the show that year was called "The Will to Motorization." Automobile industrialists pricked up their ears: the demand-inhibiting taxes on cylinder capacity and fuel really were going to fall. Street builders sniffed the morning air: someone finally wanted to support their plans for opening Germany up for a breakthrough. Hard cash beckoned to the unemployed: they would have earnings to bring home again. And the broad public sensed that everything would get moving again, including Germany's long-awaited motorization. Hitler was the first chief of state to welcome the vision of the German people rolling on wheels and the German land covered with streets. In the history of the rise of the automobile, the Third Reich plays a key role: a project that had accumulated such power of attraction over decades became a program, and as a program it was to persist long after the Reich itself had disappeared in rubble and ashes.

Pyramids of the Reich

September 23, 1933—Adolf Hitler, in his brightly polished knee boots, climbs atop a small hill and begins shoveling like the devil. The groundbreaking ceremony for this construction project, unprecedented in scale, was masterfully staged. "No symbolic groundbreaking, that was real work in the dirt," the National Socialist periodical *Die Straße* (The Road) reported, noting "the first beads of sweat" on the Führer's forehead. All the radios in the country carried Hitler's call: "Begin! German workers to the job!" Within a year of this signal, ninety thousand workers and engineers will have gotten to work rooting up the German soil.

The earth-moving could begin so resolutely—just three months after the "Reich Automobile Law" had absolved the states of responsibility, concentrating it instead at the national level—only because the plans had been ready for years and could now assist the National So-

1. *Die Straße*, 1939, no. 20, 242.

Left: *Cover of the ordinances for HAFRABA (Association for Preparation of the Frankfurt-Basel Hanseatic Highway) (1929)*. Right: *Groundbreaking of HAFRABA, September 23, 1933*. From Die Straße, *1934, no. 20, 259*.

cialists in achieving their breakthrough. The Association for Preparation of the "Frankfurt-Basel Hanseatic Highway" (HAFRABA) had introduced the idea of an "autos-only" road as early as 1925; it was promoted primarily by the construction industry and had been developed to the stage of grid plans covering the entire country. "Automobile traffic finds its primary impediments in the countryside," a memorandum of the time pointed out,

> where it is expected to share the streets with horse-drawn carriages, bicyclists, and pedestrians. . . . The modern concept of traffic engineering is to introduce a network of special highways, to serve the needs of long-distance travelers and to be used by the fastest automobiles (for which it is meant) with the greatest possible safety and the lowest possible operating costs.[2]

Without "right of way," the automobile was worth only half as much. A totally new type of road was required to make space penetrable, and

2. Quoted in Kurt Kaftan, *Der Kampf um die Autobahnen. Geschichte und Entwicklung des Autobahngedankens in Deutschland von 1907–1935 unter besonderer Berücksichtigung ähnlicher Pläne und Bestrebungen im übrigen Europa* (The fight over the Autobahns: History and development of the Autobahn concept in Germany from 1907 to 1935 with particular reference to similar plans and endeavors in Europe) (Berlin, 1955), 21.

that was an "autos-only" road, designed for high speeds, divided into separate roadways, bypassing clogged central districts, and reserved solely for motor traffic.

So thought the planners, anyway, in their bold proposals of the 1920s; but their contemporaries, not yet so unreserved in their belief in the automobile, proved skeptical. Is it all really necessary? The Weimar administration more than anyone else asked this question, meanwhile pointing to the established rail network and deciding to expand existing streets. Long-distance traffic, moreover, was little in evidence, a circumstance that prompted a memorandum expressing the following conclusion: "Automobile streets without automobiles and without a broad, upwardly mobile middle class are no more productive than a pile of gravel."[3] Just as there could be no question of a pressing need for freeways, in terms of the economy of transport a new class of streets was equally senseless. It was purely and simply the auto-industrial complex, including street planners and construction and cement interests, who obstinately pursued the plan of boring holes in mountains and crossing valleys with bridges to forge a place in society for the automobile.

Despite various lobbying intrigues, the matter remained undecided until the eve of the Nazi seizure of power. But then the times changed. Hitler viewed the massive highway construction project as a key investment that, in addition to stimulating automobile production, would give the German economy a boost. Certainly jobs would be created; construction workers were once again as busy as the steelworkers, and now the motor of trade could start up for a whole group of industries. With some success: nearly a million jobs (of which some 120,000 were linked directly to automobile production) can be traced to impulses initiated by the motorization policy. In the forefront was the production value of motorization—"rolling wheels on country roads mean rolling wheels in the workshops"—while direct military plans played a minor role up until 1937.

The proposals of traffic engineers from the 1920s, however, were only a framework around which the National Socialist image of society could be wrapped like a cloak. Nevertheless, the brown movement inspired fascination thanks to its attempt to give technological transformation—which, Janus-like, had in the Weimar period caused

3. Quoted in ibid., 130.

Postage stamps of Third Reich highways. From Die Straße,
1936, 664.

unrest and anxiety—a home within a vision of nationalistic gran-
diosity. Gone were the layoffs and humiliation; the financial vultures
and bleeders of the economy were no more. Cold industrial progress
seemed to have been plunged into the warmth of the nationalistic
worldview. Under the title "National Highways—The Sign of Our
Times," the following appeared in the pages of *Die Straße:*

> The contradiction that existed between the technical development of
> the motor and the very limited realities of the street—a result of insuf-
> ficient attention in previous decades—is now being cast aside. The
> roads of the Führer will be developed into great traffic arteries, which
> not only will contribute to the melding of the German people into a
> stronger political and economic unit, but will also put an end to the
> last remnants of particularistic thinking.[4]

The automobile, braked for so long, finally gained right of way.
Now traffic could pulse through space from border to border unhin-
dered by bottlenecks, and the most distant nook could be connected to
the "circulatory system" of national life. The bright strips of highway,
running like "arteries" through the body of the nation, bind north and

4. *Die Straße,* 1936, no. 1, 8f.

On June 23, the law on the construction of national highways will be proclaimed. The world will receive it in wonder. It represents the grandest road-planning project of all time. History provided no model for this undertaking; all known street constructions of the ancient Romans and Teutons, all road projects of modern peoples, of Napoleon or whoever else might have served as a model, they all pale in the face of the power and breadth of this plan. Not only in the history of the motor vehicle, in the history of transport altogether, but also in the history of street construction, a new chapter will be opened on June 23, 1933.

Seven thousand kilometers of national highway are to be built, for no other purpose than to serve motor-vehicular traffic. They are to be entirely free of intersections, free of danger, without oncoming traffic—colossal traffic arteries that extend through the entire country, bringing the most distant cities nearer one another in a way never known before, allowing traffic to proceed at speeds no one ever dared hope for. Nothing is to cramp or delay you in your swing from one horizon to the other, no tollgates and no intersections, no oncoming automobiles and no unmotorized traffic. No twisting trips through localities will disturb your smooth progress and riddle it with danger spots. As flawless and powerful as the National Socialist revolution itself, the highways will spread across the land.

But they are to be more than the grandest, best, least dangerous, fastest, and most modern roads on earth—they are also to be the most beautiful roads in the world, the noblest adornment to the noble German landscapes, which in this conception are to sparkle like a stone in an artfully wrought ring. They are not to be a destruction of nature, the ugly product of a degenerate technology, but the supreme crowning of the landscape, a crowning that the earth always receives when the human spirit is married to its natural beauty, thus intensifying it.

It is not roads, then, that are to arise, but works of art—just as once temples were built, and not huts; cathedrals, and not mere prayer houses; pyramids, and not tombstones. That is what the Führer wants.

For the first time in the history of humanity, the Führer is elevating the street above the domain of the natural path and artificial road construction, and into the sphere of art.

From W. Bade, *Das Auto erobert die Welt. Biographie des Kraftwagens* (The automobile conquers the world: Biography of the motor car) (Berlin, 1938), 316f.

The model highway. This photograph from Irschenberg makes use of compositional elements common in the history of European landscape painting.

south, east and west; driven by the heartbeat of the center, they allow common thinking, feeling, and doing to circulate in the form of information and commodities.

The slogan "One People, One Reich, One Führer" did not come about by itself; out-of-the-way places and resisting groups had to be forced into unity. The *Volksgemeinschaft* (national society) could only be built once the body of society was cleansed of obstructionist groupings (the unions), contrary intentions (the Communists), and inaccessible areas (conservative rural communities). The erasure of stubborn differences was supposed to be effected by "eliminating opposition," and the highways were the spatial expression of this venture. "To meld the German people into unity"—this social vision was to be set in the concrete of the highway network: the vision of a homogeneous society in which the pulse of life beats to a uniform rhythm, unopposed by local consciousness or cultural particularity. This vision was (and is) modern in the highest degree, and one should not be fooled by the rhetoric of "blood and soil." For the "circulation" metaphor had been in vogue since the nineteenth century whenever adjusting the inner life of society to the heartbeat of the modern was at issue. Money, goods, information—they all "circulate"; and Baron Haussmann constructed his boulevards through old Paris to open the city to the circulation of air, water, and goods. The fascists simply gave a nationalistic interpretation to what elsewhere was accomplished under the name of the "market" or the "plan." Whether in orders, wares, or norms, industrialism can become fully effective only when it coincides with the creation of a homogeneous society.

The National Socialists were nevertheless scarcely inclined to justify freeways in the language of liberalism and reorient themselves to the free circulation of commodities or, surely, thoughts. They had, after all, taken the stage to fight the "disintegrative" spirit of liberalism and to replace the ties that progress was dissolving with a totalitarian version of nationalism. Handicraft workers and small business people, obedient heads of households, and amateur politicians enamored of order—in short, the petty bourgeoisie, supporters, at least in the beginning, of the National Socialist German Workers' Party (NSDAP)—they especially, confused and beset by the collapse of the old world following the shock of industrialization, longed for a strong state to return to them their lost security. Their hearts pounded as the Führer's scratchy voice promised them a breakthrough into the past, to homeland, fatherland, and the company of the select among nations.

Nevertheless, the promises of progress were alive in their hopes even as it threatened them; they dreamed of a progress of harmony and order, the reconciliation of reaction and modernity. And this was what the National Socialists promised. The party presented the greatest construction project in German history—6,500 kilometers of road planned, 3,500 kilometers actually built—not only as a technical accomplishment of the highest order, but above all as a cultural feat to the eternal glory of the nation. The inspector general of German roads, Fritz Todt, could barely contain himself as he celebrated the first thousand kilometers:

> It is once again a matter of pride to be a road builder. . . . The German Reich is getting roads the likes of which, in their magnitude and beauty, have never been built in the history of human culture. . . . In

the conception of National Socialist technology, roads are works of art, just like architectural structures. . . . From this will to culture results as well our relationship to the landscape; therefore we fit the roads into the landscape with care for the trees and shrubs![5]

Monuments for the future, pyramids for the Reich—this is what freeways were supposed to be, not soulless strips of concrete. The much sung German landscape was to be enhanced, not destroyed. What would dominate was not the straight line, but the gentle curve; not the throughway, but the beautiful view. The hilltop of Irschenberg, on the freeway from Munich to Salzburg, will serve as an example: an incline, uncalled for by geography, but directed upward to provide a special vantage on the tidy hills and onion domes before the backdrop of the towering peaks of the Wendelstein. To create not the shortest but the noblest connection, that was the motto (whatever consolation it might offer drivers of today precariously traversing the incline in icy weather). The monumental "work," the system of freeways was conceived as a reconciliation of that which many considered had been torn asunder: technology and culture, machine and Valhalla. From this ambition, in a rather backward nation, emerged what later became the model even for more advanced countries: the most efficient long-distance highway system of the time.

Every Month in Marks Save Five,
If Your Own Car You Want to Drive

Not a single privately owned Volkswagen ever rolled on the streets of the Third Reich. And yet this "car for everyone" loomed on the horizon of the National Socialist future, a future for which working and perhaps suffering appeared worthwhile because it promised that one day even the little man would partake of the blessings of technology. Meanwhile, only the big words came cheap; but these words hit dead center the emotions that had built up for the automobile over the past decades. "As long as the automobile remains a means of transport for especially privileged circles," Hitler declared at the 1934 Berlin Automobile Show,

> it is with a bitter feeling that millions of obedient, diligent, and able fellows, who in any case live lives of limited opportunities, know themselves to be denied a mode of transportation that would open for

5. Ibid., no. 19.

'Every week in marks save five / If your own car you want to drive." Three hundred thirty thousand people saved five marks a week—and never saw their "Power Through Joy" cars. (Courtesy Volkswagen AG, Wolfsburg)

them, especially on Sundays and holidays, a source of unknown, joyous happiness. . . . The class-emphasizing and therefore socially divisive character that has been attached to the automobile must be removed; the car must not remain an object of luxury but must become an object of use![6]

How wonderful such words must have sounded to those who had so often counted their money, in vain, to see whether it might suffice for a "Tree Frog"—used, naturally. Suddenly the highest authority had decreed that all the hopes that had built up in the Weimar years

6. *Völkischer Beobachter*, March 9, 1934.

Cover of Motorshow *(1938). The* Volksgenosse *on the go.*

could flow freely: the automobile, the symbol of modernity and elegance, might be brought within the reach of all; everyone now had the prospect of perhaps one day being able to drive a small car home; no longer would Germans have to glance enviously at America, where Ford bestowed on the land a genuine flood of automobiles. While in the twenties the automobile had conquered the desires of the masses, recreating a tense distance between those at the top and those at the bottom, the National Socialists now pledged to overcome the distance and help even the German worker get some wheels. Again the same motto: Yes to technological progress, not only for the elite, those snobs and swindlers, but in the name of the modest but righteous happiness of people like you and me. It had been painful for the "obedient, diligent, and able fellow" to look on as others basked in the sun of progress, while all his drudgery ever got him was margarine and a damp apartment. The regime lent recognition to the aspirations of the petty bourgeoisie in that it promised to strip the automobile of its luxury status in favor of a solid status as a use object.

Hitler was virtually obliged to promote the Volkswagen project, because any stimulation of automobile production had to do more than merely further the motorization of more prosperous groups. Such a development would have been "un-German," for the figurehead of the movement was, after all, the *Volksgenosse* (fellow German)—and how could the automobile be withheld from him? Moreover, a broad program of modernization suited Hitler's purposes perfectly, in that it gained the support of the masses for a new social model, one that he, in his genius, gave to the people and one that loosened the received loyalties—to family, church, the traditional milieu, and all the old authorities—as it created a direct connection between himself and all *Volksgenossen*.

Hitler needed modernity, however little it appealed to him. The First World War proved to be the great crucible of modernization. Afterward everything tended toward mass society, in which people no longer behaved as befit old-fashioned tradesmen, blacksmiths, or even counts, but in accord with their interests as wage earners and consumers. In that people responded increasingly to the same signals, whether prices, broadcast news, or advertisements, they also tended to become more similar and predictable in their reactions. The common signal devalued the word of the father, pastor, or neighbor.

The masses are a consequence of the centralization of social life, a fact that was opportune for fascism as well. Indeed, fascism is hardly conceivable without this development, for the "Führer principle" was to command the whole of society and subvert all other loyalties. The *Volksgenosse* is the mass individual in fascist dress; that is why the National Socialists fortified the tendencies toward a production and consumer society. To aid the "integration of all Germans in the process of national production" came a supply of standardized consumer goods for the *Volksgemeinschaft:* one thinks here of the people's radio (*Volksempfänger*), apartment construction—and the Volkswagen. Once again: industrialization, based on the separation of producers and consumers, offered the framework for the fascist social project. But not only that. Just as for Lenin Soviet power plus electrification was the formula for communism, so Hitler saw National Socialism as arising from the *Volksgemeinschaft* plus motorization.

Hitler, an admirer of Henry Ford, viewed the Volkswagen project in the context of (antiluxury) mass consumption, as his reference to the success of the people's radio points out:

> Just a few months ago German industry, through the manufacture of a new people's radio, succeeded in delivering to the market and selling an enormous number of receivers. I wish now to place before the German motor vehicle industry the significant task of designing an automobile that necessarily comes to attract buyers numbering in the millions.[7]

The man who took on that task, Ferdinand Porsche, also referred to the people's radio in a memorandum on the subject. Porsche, already famous as a designer of race cars, had toyed with the idea of a small automobile since the early twenties. He had drawn up a series of proposals but had not managed to overcome the disinterest of the industry. After his break with Gottlieb Daimler he had set up a design office of his own in Stuttgart, and there he became so involved in the idea of a people's automobile that even the Russians had taken note. They gave him a tour of Russia and proposed that he adjust his plans to suit the Soviet people; this idea, however, he refused, primarily for personal reasons. Porsche, the classic unpolitical technician, finally saw his time arrive when Hitler, working against the delay tactics of the automobile industry and bypassing the responsible departments, called on him to design the car that he, Hitler, had in mind: a vehicle that was economical to operate, with four seats for the whole family, air-cooled to avoid the problem of freezing winter temperatures, capable of a

7. Ibid.

*from the meeting on April 11, 1934, concerning the
creation of a popular automobile* [Volkswagen].

Represented were the Reich Chancellery, the Reich Ministry of Education and Propaganda, the Reich Ministry for the Economy, and the Association of the Reich Automotive Industry. Hon. Dr. Brandenburg (D.E.) acted as chair.

The chairman referred to the speech by the Reich Chancellor on the occasion of this year's International Automobile and Motorcycle Show. He reported that the Führer has charged the automotive industry with the goal of creating a Volkswagen. In order to fulfill the assigned task, clarity must be achieved as to the nature of the future Volkswagen and the standards to be set for it. The standard price must not exceed 1,000 Reichsmarks, and the operating costs must not exceed 6 Reichspfennigs per kilometer. Nevertheless, the automobile must be operationally sound and accommodate three adults and one child.

The expert representative of the Reich Ministry for Traffic then discussed the design possibilities for construction of such a car. One feasible solution appeared to be the three-wheeled design, with two wheels in front, one wheel in back, and a rear engine. The advantages to be noted: symmetrical drive, good cross-country mobility, low rolling losses compared to a four-wheeled car, low weight, and a naturally streamlined shape.

The representative of the Association of the Reich Automotive Industry declared that the industry has already taken up the matter. Because, however, the opinions of the designers diverged widely, the question did not yet admit of a conclusive answer. And one must not neglect consideration of the risk to the industry. It would be a serious mistake to give the industry design requirements. The solution to the problem must be

highway cruising speed of 100 kilometers an hour, and, above all, costing no more than 1,000 Reichsmarks. A demanding, indeed nearly impossible assignment. Yet the first thirty test models were on the road in 1938 and presented to the excited public in the same year.

"Simple, modest, and reliable"—and so it was, just right for the "millions of obedient, diligent, and able fellows." The automobile had finally cast off its luxury character, so that nothing of the status carriage remained. The "KdF-Car," as the Volkswagen was still called—for *Kraft durch Freude,* or "Power through Joy"—was in its design a full-fledged automobile in its own right, no longer a large car thinned down nearly to the point of uselessness, as the small cars of the Weimar period had been. It is understandable that Wilfried Bade rejoiced in his propagandistic history of the automobile: "Until now the automobile left largely to the industry. On request, proposals could be submitted to the government, and this could indeed transpire quickly, if desired by May 15 of this year.

The representative of the Reich Chancellery confirmed the presentation of the meeting's chairman. He said that the industry builds too many expensive cars, which do not correspond to the distribution of income among the broad social strata. The price of the Volkswagen must not be allowed to exceed 1,000 Reichsmarks. If necessary, the risk to the industry could be minimized through support from the government.

The representative of the Reich Ministry of Education and Propaganda characterized the instigation of the Reich Ministry for Traffic for the production of a three-wheeled Volkswagen as absolutely valuable. The Volkswagen must be inexpensive to purchase and to operate. For such an inexpensive car there would also be the possibility of export.

The representative of the Reich Ministry for the Economy recognized the risk to the industry. The industry, however, has as yet built no first-rate small vehicle. A market opportunity must exist. Only by means of mass production could the inexpensive Volkswagen be produced.

The liaison officer from the Reich Ministry for Traffic detailed the standards to be set for a Volkswagen from his point of view.

The result of the meeting was to define the character of the Volkswagen according to the following conditions:

Seats for	5 adults and 1 child
Fuel consumption per 100 km	to 5 liters
Maximum speed	80 km/h

Cross-country mobility and ground clearance corresponding to a powerful motorcycle.

Prepared in the Reich Chancellery; no signature or initials.

has conquered the world. Now begins the true possession of the automobile by people."[8] In its outward appearance alone it expressed the idea to which it had given material form: the *Volksgenosse* in motion. Its solid construction, its economical look, its down-to-earth simplicity gave eloquent testimony as to its target market. There was no elongated hood, no limousine-like coachwork, no elevated passenger compartment—no, it appeared sooner snub-nosed, something one descended into rather than mounted, and it stooped humbly on the street. Truly a classless vehicle, a vehicle for the people.

Even so, its design offered a model representation of the dominant

8. Wilfried Bade, *Das Automobil erobert die Welt. Biographie des Kraftwagens* (The automobile conquers the world: Biography of the motor car) (Berlin, 1938), 356.

formal language and was wholly in the spirit of the time: the stream-lined shape symbolized speed, the predominance of diagonals suggested motion even before it moved. The streamlined form marked automobile design of the thirties: vehicles were lower, with roof lines made to glide under the wind, just like the angled front and the fastback. Body sections were joined together in a unified sheet metal hull, whose smooth surfaces and gentle curves recalled a bird soaring through the air or a fish swimming through water. Automobiles seemed to cut through the air, promising easy passage; it was not by accident that a Mercedes "Freeway Courier" and an Adler "Freeway Car" stood by, ready-made for the Führer's streets. The streamlined look characterized both the status vehicle and the small car of the time; whatever the distance yet dividing them, in their forward impulse they were one. Even beyond the automobile, streamlining set the style: radios, irons, even pencil sharpeners appeared in racy forms. The race car drivers Rudolf Caracciola and Bernd Rosemeyer, rushing from one triumph to the next, were, after all, the heroes of the day, and Hans Stuck acted in a film with the title "Full Throttle to Happiness."

This happiness beckoned, and for only 5 marks a month one could claim it for oneself. Beginning in 1938, the German workers' leisure organization Power Through Joy offered a savings plan by which four years and seven months' worth of diligent discount stamp pasting was supposed to reserve a Volkswagen in one's own name. Was supposed to—for the 336,668 savers, who together paid 280 million marks into the fund, never saw their Volkswagens, even though by 1939 the gigantic automobile factory in present-day Wolfsburg had already been conjured up from bare earth, ready to start production. (Hitler, in 1937, said: "There can only be one Volkswagen in Germany, not ten.") But from 1940 on, the assembly lines at Wolfsburg were producing jeeps, the Volkswagen's military brother. And it wasn't in leisure clothes, but in uniform, that Germans set off for those lands they had always wanted to visit—only as vacationers. A people on wheels—the vision suddenly took on military features. The private "auto-mobilization" turned into the general mobilization for war.

Disappointed savers continued their lawsuit against the VW factory until 1961.

We Did It!
(1950–1960)

"If someone," wrote Kasimir Edschmid in his notebook in 1960,

> in 1948 had told one of the millions of worms now driving around in
> their cars that he would one day be prosperous, well equipped, off on
> foreign vacations with good money in his pocket, esteemed, and alto-
> gether a respectable human being (this just after he had gathered the
> cigarette butts of occupation soldiers off the street), he would have
> rolled his eyes and pronounced the prophet an imbecile.[1]

These years were like a miracle: rubble was transformed into glass pal-
aces and bomb craters into boulevards; refugees became homeowners,
and economic wizards were made of black marketeers. A page in the
history of nations has probably only rarely read as it does for these ten
years in Germany. Where hunger had just been the rule, now one
fought as a heavyweight; and where pilfering coal had been commonplace, now purchasing power got all the attention. The fifties were re-
markable not for the contrast they offered to the Nazi period, but for
their contrast to wartime. It was the liberation not from injustice but
from chaos that released the outpouring of atonement that in turn fed
the optimism of reconstruction. The chaos also offered advantages:

1. Quoted by Hermann Glaser in *Das Wirtschaftswunder: unser Weg in den Wohlstand,* ed. Frank Grube and Gerhard Richter (Hamburg, 1983), 181.

one could now make a clean sweep of the past—always an obstacle to progress—and stride toward completion of the consumer society, which, having taken form in the twenties, was pushed in the thirties to the far horizons of mass expectations and finally now arose as a reality from the rubble of the old world. Apartment, car, travel—the wishes of the twenties, mere promises in the prewar period, now established themselves as needs, only to be transformed yet again in the sixties into expectations. One donned the sackcloth and rolled up the sleeves, repressed the guilty conscience, and plunged into reconstruction, showing remorse and cleansing the soul by joining the West in its "American way of life." A happy arrangement, in truth, for the Germans proved themselves able to learn, able to get on with making for themselves what they had always envied in the big brother across the pond and what in the confusion of Weimar and the catastrophe of war had been lost track of: the modern society built on consumption and commerce.

The Miracle at the Door

Nineteen fifty: with the dust of the collapse just settled, the automobile embodied most palpably the change in history's course. The dark times were past, privation was still fresh in the memory, and a whole people proved nearly unanimous in its deep resolve: away with yesterday, life begins anew!

> Freedom calls, freedom presses . . . ! Can anyone hold it against people if they, who have waited eleven years for this moment, can, on a wonderful spring morning, get the car out of the garage, pack it full, and then charge off in it? With what desire and what wanderlust this moment has been envisioned during the past eleven years. . . . In these eleven years so horribly stolen from us. Eleven years in which we've bypassed real life—as soldiers, or as the ones who stayed at home to be threatened by the bombs. Eleven utterly difficult years; for some of us, they should have been the best years of our lives. Should we not attempt to recapture the lost time? Should we not hurry, rush to experience at least a tiny remnant of our "good" years?[2]

All the senses were now devoted to shaking off the past, to striving, after all the collective unhappiness, for private happiness, and to living right again. Ideals were not in demand, or politics. People wanted to resume the "normal" life of days before the war, resume the small

2. *Das Auto* 9 (1950): 289.

The Christmas child of 1950.

pleasures of leisure and consumption brought by those first, "reasonable" years of the Nazi regime. Freedom: that meant to have escaped the mill of fate and to be able, finally, to enjoy a little happiness; it meant to leave misery and humiliation behind and sniff the fresh air of the future. After all the marching in step and living in barracks, it was not surprising that people were eager for freedom in their private nooks; after the fires and bombs it was natural to seek freedom in the possession of colorful goods. For a sense of freedom such as this, the automobile was a powerful symbol.

> We did it: the new car is in the drive. All the neighbors are peering out their windows and can see that we're preparing for a little weekend excursion. Yes sir, we've done something for ourselves, we want something from life. After all, that's why we're both working, my husband in the plant and I as a secretary for my old company.

This advertisement for Ford Taunus shows father in the foreground, in nylon shirt and tie, putting a suitcase in the trunk. His wife, in a checkered dress and sporting a permanent wave, looks on with satisfaction as she holds the hand of their small son—in neatly pressed trousers and wearing a bow tie—while the neighbors stick their heads out the windows of the plain, newly constructed apartment complex.

The scene is tellingly illustrative of the social dynamic of the 1950s. A triangular relationship is depicted: the consumer good (the car), the owners, and the others. The owners, the nylon father and the checkered mother, do not want merely to live well, they "want something from life." The language alone makes clear that for them the good life is equivalent to having, to accumulating goods. Happiness dwells without, beyond one's own person, in a world of things to which one must adapt if one is to partake of the good life. Only a car can create this freedom. Things, however, are in short supply, for they carry a price tag; one must therefore "do something" for oneself. That is why our happy owners work, he "in the plant" and she for her "old company"—so they can save up purchasing power. In effect, the good life is reduced to purchasing power: those lacking a fat purse have to look on, able only to peer out their windows and cast envious looks—in which, completing the circle, the owners get to bask. Avarice, conspicuous consumption, and envy—by this threefold formula the German people gradually transformed themselves, stratum by stratum, commodity by commodity, into a consumer society. And the automobile assumed a leading role in the change.

For the next twenty years—until around 1975, when one could speak to a certain extent of full motorization—the envy was granted an ample chance for expression. By 1962 a mere 27.3 percent of German households had their miracle parked in the drive; throughout the fifties the Lloyd advertisement, "Productive in cheer, happy in pleasure," was valid only for the upper reaches of society. But in the course of the sixties, because earning power went up while prices went down, more and more people not only could afford to fulfill their dream, but also saw their assumed right to prosperity redeemed in the possession of an automobile. This was exactly what Ludwig Erhard promised, in his 1957 book *Wohlstand für alle* (Prosperity for everyone), as the reward of the daily grind: the democratic consumer society, equality before the commodity. What the man with the mustache and cowlick only dangled before the eyes of the *Volksgenosse,* the fat man with the

"Where building is under way. . . . Where precise calculations help reduce personal costs—in industry and trade as well—where higher profits without increased expenditures are the goal—that's where you can't do without a LLOYD. Ten marks a month in maintenance costs at barely 5 liters fuel consumption per 100 kilometers for a LLOYD passenger car or station wagon. That means traveling economically. That means transporting cheap. DRIVING A LLOYD MEANS STAYING IN THE BLACK."
Advertisement from Der Spiegel, *1952, no. 45.*

cigar wanted to make into a concrete reality for the consumer in mass society—namely, a car in the drive.

And the offering expanded like a Japanese fan, from the Opel Kapitän to the Messerschmitt Kabinenroller, bringing cars within the reach not only of rich manufacturers but of poor mail carriers too. The compact and mini cars of the time provide a nearly physical measure of the effort made to shrink the coveted automobile to the dimensions of smaller purses. Brütsch and Lloyd, Isetta and Goggomobile, stooped as far as they could to meet the slow but steady rise in purchasing power of the masses.

"A proper personal automobile for 2,800 marks," came the excited judgment of a test driver as the Lloyd LP 300 appeared on the streets. "It's no pseudo-automobile, like most of the small cars that have come into the world of late, the kind where you can't tell whether it's a motorcycle with four wheels or an automobile with motorcycle qualities." Appreciative astonishment followed:

> The little car has a smart appearance; the so-called pontoon body makes full use of the width; the two separate front seats are comfortable enough; the back seat has plenty of room for two children (or lots

of luggage)—[though] for adults with the plumper bodies that are once again in style, it would hardly be bearable over long stretches.[3]

"The Plastic Bomber," it was called derisively, because its body was a hardwood frame covered with plywood and artificial leather. But that did not diminish affection: nearly four hundred thousand of this, the most successful of the German small cars, and its successor models had been sold by 1961.

While the Goggomobile, another of the generation of minicars, bowlegged and full of rattles, never quite lost the air of a make-believe automobile, the BMW Isetta stood out among the small cars because of its unique character. With the door in the front, so that one got in around the tiltable steering column, the motor positioned sideways behind the passenger seat, and a compressed length of only 228 centimeters, it had the appearance of an easy chair on wheels. Produced from 1955 to 1962, it was an ingenious and unpretentious piece of automotive technology, although, because it lacked a passenger buffer zone against crashes, it could not hold its own against the faster models of the sixties.

Nowhere else in the fifties did the wave of small cars achieve the crest it did in Germany, a visible expression of the wholesale public desire to invest every last dime in an automobile, even when the typical savings sufficed only for an obviously undernourished vehicle. The small car was particularly effective in converting motorcycle riders to four wheels, but it also allowed entry into the "lower middle class." At the ready in these years was the parade piece of the economic miracle, the Volkswagen. Already promised in the 1930s, it had become the symbol of betrayed expectations, of the "stolen eleven years." With its unmistakable classical form it was conspicuous evidence of the extent to which the 1950s proved to be the continuation of the Nazi period by market economic means. The dreams survived, but their fulfillment had fallen into the rubble; could an ascent from such depths be perceived other than as a miracle? Yet it ran and ran and ran some more, the Volkswagen, from the shores of Helgoland to the red sun of Capri. And it remained ever true to its character: simple but eager, economical but solid. Gray in color, plain in form, modestly equipped, it still exuded that antielitist aura that had already in the thirties won it a place in the hearts of the upright *Volksgenossen*. One

3. Ibid., 282.

German compacts of the 1950s. Left to right: *Viktoria Spatz, Messerschmitt Kabinenroller, BMW Isetta, Messerschmitt Kabinenroller, Maico, Gutbrod, Zündapp Janus. (Courtesy Vereinigte Motor-Verlage, Stuttgart)*

could afford it without being ashamed, enjoy it without coming across as a braggart or a king. No, one sought practicality at a good price, far from the shabby Lloyd but also at some distance from the pretentious Ford. Thus did the "lower middle class" establish itself, and by the mid-sixties every third driver who cruised the streets would be in a Volkswagen.

Still, those who wanted to show off their newly won prosperity found an adequate expression in the Opel Rekord or Ford Taunus. As the two American offerings in the German automobile selection, both had something of the sheen of the American dream car. They took on, in diluted form, design elements from the parvenu baroque of the street cruisers: chromed radiator grills, covered headlights, panoramic windows, rear fins, and snazzy chrome trim. Such ornamentation, preferably in company with whitewall tires, captured something of the American glitter and suited to a tee the nouveaux riches who had come into quick money in the young republic. The upright Volkswagen was not their thing; but then again, the self-possessed luxury of the established elites often seemed to them too old-fashioned. They wanted to appear in sync with the new, the dynamic times. Although the top flight of German automobiles—Mercedes, Opel Kapitän, BMW limousine— was not altogether closed to these individuals, the look of these cars was still predominately along the classical lines of the thirties, with the prominently peaked hood, the swept fenders, and smooth, rounded surfaces, expressing a genuine elegance and a tie to tradition that abhorred all presumption. Moreover, the outward appearance of both BMW and Mercedes carried strong associations with the prewar period. These cars were therefore perfectly suited to demonstrating the continuity of the leading social groups. Finally, the spirit of such cars was oriented toward tradition; conveying little of the American way of life, they demonstrated a value placed, in the manner of Konrad Adenauer and Cardinal Frings, on authority and bourgeois order.

We're Something Again!

With bouquets on the hood and a license plate reading "5 Million," the Jubilee Volkswagen (otherwise every bit the trademark car) rolled off the assembly line in December 1961 toward the cameras, lights, microphones, and honored guests. After the applause and the popping of corks, Heinrich Nordhoff rose for the keynote speech. To mark the significance of the day with the appropriate context, he cast a glance at the dismal year 1948: "Seven thousand workers were painfully producing at the rate of a mere six thousand cars a year—provided it did not rain too much. Most of the roof and all the windows of the factory had been destroyed. Pools of stagnant water were under foot." Then he led his listeners through the intervening thirteen years, the pathos in his voice rising as he neared the conclusion: "We have good reason to be proud, [for the Volkswagen] is a symbol of one of the biggest industrial successes ever reached . . . , the result of hard, unremitting work and of diligent attention to a correctly set goal."[4] The success of the Volkswagen plant was also a national success: the Bavarian mint took the occasion to issue gold and silver medallions, a Volkswagen on one side, Nordhoff's profile on the other.

The miracle in the drive corresponded to the miracle in Wolfsburg. To the exhilaration of consumption, on the one hand, belonged the glorification of production on the other. This was especially true in postwar Germany, of course, where the rapid ascent from rubble and ashes struck many as a resurrection; stylish furniture, no-iron shirts, and new apartments made the apparatus of industrial production appear a lifesaver. Now that the old formulas—"fatherland," "nation of culture," and "major power"—lay entombed on the field of honor, the nation (now, moreover, divided) had to find meaning in a new social

4. Quoted in Walter H. Nelson, *Small Wonder: The Amazing Story of the Volkswagen,* rev. ed. (Boston: Little, Brown, 1967), 131, 155.

model: as an economic power, producing consumer happiness. The upward curve in production statistics came to bespeak movement upward toward the greatest happiness for the greatest number, and the object of social life was taken to be increased consumer happiness— these assumptions became in the 1950s articles of faith, the dogmatic nature of which was not recognized because their truth seemed to be confirmed ever anew in the repetitions of daily life. Radio broadcasters focused their reports on weekly coal extraction figures; factory dedications, enterprise launchings, and export figures made headline news. Production gains were celebrated much as successes at the front had been—for output of goods promised prosperity for everyone.

It is no wonder, then, that Volkswagen could get by with a single type of advertising until 1961: the VW production statistics, published as a triumphant graphic in the newspapers at year's end. Take, for example, the advertising illustration for 1953: a ribbon spirals dynamically upward; a VW, winding its way to the top, is just reaching the highest point; the ascension as a whole is titled, "More and more, better and better." Constantly upward, from the depths into the heights, as natural as a spiral, more products every year—could a belief in growth be illustrated more graphically? The time was guileless in its optimistic trust in progress, innocent in its overestimation of itself. Hence the ad's resonance.

Everywhere it appeared that the easy life was found in the burgeoning stream of goods; just press a button and the world came fluttering into the house on television, twist a knob and the central heating started up, a little pressure on the gas pedal and the car sped away. The possibilities seemed inexhaustible, assuming only assiduous productivity: one day the atomic reactor would surely be domesticated for the kitchen stove, and the Sahara for orange trees. The world offered endless resources, and technology coupled with ability seemed to reclaim the old dreams; what difference, then, if in 1960 it took the average West German only a day to use up the three liters of oil that in 1950 he consumed in a week?

Perhaps the military and moral defeat could be squared with technology and ability as well: Germans had been known for technological ingenuity, after all, since the time of Werner von Siemens and Rudolf Bosch. Greatness thanks to export power—one could seize this line of national self-understanding, cause the guilt to be forgotten, and once again win prestige in the world. The annual reports from VW took up

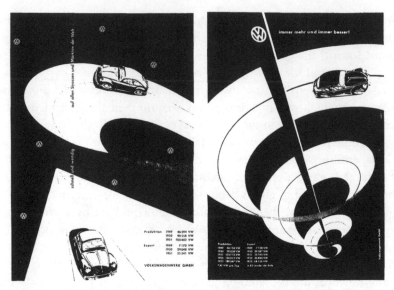

VW annual reports: "Fast and maneuverable on all streets and in all markets. . . . More and more, better and better!"

"Examined from all sides . . . the sensible automobile. . . . Volkswagen, ahead by far."

Heinrich Nordhoff, Volkswagen general director, and his Beetle.

this motif, praising the product as "quick and nimble on the streets and in the world marketplace" (1951); as "desired the world over" (1952); or simply as "the worldwide success" (1963). And, along with the five million cars produced in 1951, the company's world status was celebrated as well, for approximately half of this number were sold abroad, smoothing the way for the Bug to become the most successful automobile in history (twenty million sold by 1981). Besides ringing coins, export strength brought self-confidence home to Germany, and with trade statistics the nation worked its way to being the model student among the Western powers. On a Wolfsburg city map of the time appears the candid remark: "Here [in Wolfsburg] one of the German miracles, the Volkswagen, was created, a product of the German creative spirit and diligence that has found recognition and inspired admiration the world over." Thus did the automobile become the symbol of an epoch, the palpable sign of the economic potency of a morally humiliated nation that, in defiant relief, offered itself as proof of Chancellor Erhard's call: "We're something again!"

Republic on Wheels
(1962 – 1973)

It is hard to believe, but only twenty years have passed since motorization began rolling across the country. It was as if a dam had broken: in the thirteen years from 1960 to 1973 the number of private automobiles quadrupled, the number of kilometers driven tripled, and the total extent of freeways doubled. Historic sections of cities were bisected by thoroughfares, freeway interchanges cut into forests, and, whether on vacation to Spain or down the street to buy some cigarettes, people took to their cars under the careful watch of traffic reporters and emergency helicopters. The wheeled society was launched.

Everywhere the onset of the sixties coincided with the emergence of the modern world from behind the old-fashioned facade of the fifties, from behind the propriety and authority that, despite nail polish and rock 'n' roll, despite oil refineries and atomic research, had set the image of the postwar society. It was no accident that the young republic had been grateful for the rule of a patriarch in the Schaumburg Palace. Konrad Adenauer's stiff dignity had given voice to a patriarchal-authoritarian conception of society, one that maintained the citizen in the role of subordinate and the consumer in the role of upright customer. Varied social groups were driven by a sense of petty bourgeois propriety, which, while it did bring all kinds of consumer goods into the household, was incapable of producing a dynamic and versatile society. The climate of the fifties was therefore strangely rigid:

yes to lipstick, but not too provocative; yes to Hollywood, but with a "clean screen"; yes to education, but preferably in parochial schools. Everywhere there was opposition to the commercialization of life; a consciousness of duty and morality stubbornly persisted, avoiding the pull of the expanding market society. The attempts to smooth the edges of the coming automobile society were almost touching: St. Christopher medals called on the protection of heaven; the "chivalry at the wheel" movement was joined to salvage the old virtues amid the bustle of traffic.

During the sixties these inertial forces declined, making way for a market- and supply-intensive society. Adenauer no longer held his own next to Kennedy. Gone was the musty atmosphere of the fifties: from now on the fresh air of rationality and democracy would rule. Equal opportunity, global control, and spatial planning were the catchwords, holding out the prospect of modern social relations. As the authorities yielded to the planning staff, so did the citizen's sense of propriety give way to the consumer's desire for goods. Even in the most distant villages, the promises of consumption now fell on fertile soil—fertile because purchasing power was on the rise, but fertile too because frugality and skepticism about progress were played out. Shaking off the conservatism of the Adenauer era, the state now concentrated on controlling the swelling stream of traffic through planning offices and expansion plans. On the other side, mass consumer demand for automobiles, building for a long time but never satisfied, now burst loose.

The Great Chase Is On

"Three years of hard work—the most modern automobile factory in the world is now a reality": thus did the Adam Opel Company present itself in a full-page advertisement in the fall of 1962. "Opel built it in Bochum. A gigantic plant—for only a single automobile model." It was worth the proud sum of 1.2 billion marks in investment capital for Opel to defy Volkswagen with a new car: the Opel Kadett. "Bochum" sounded the charge against "Wolfsburg": with one thousand cars slated to roll from the assembly line daily, the race was on for sales in an exploding market. The periodical *Auto-Motor-Sport* saw it exactly this way, commenting: "The great chase is on: Opel Kadett." It continued:

> In this light, the little man can do no better than invest his money in
> automobile stocks. But the burned child in the little man fears the
> fire. . . . The automobile giants, General Motors and Ford, however,

1. *The automobile is a commodity for everyone* for the satisfaction of daily needs, as befits the progressive shaping of our lives in a free world.

The years in which the automobile was an attribute of wealth and luxury are long since past. They ended when farsighted men like Henry Ford and others, recognizing its general usefulness for individuals and for society, made the motor vehicle into a mass product and therefore a commodity for everyone.

2. Social and economic policy in all countries has acknowledged this fact. It regards motor transport not only as the engine of a modern economy, but in the same way it regards the private *possession and operation of an automobile* as a share in prosperity, meant for and open to every citizen. The further increase in motorization is for that reason not only a necessary consequence of growing "prosperity for all," but also an explicitly declared political goal of the state.

In all programmatic declarations of leading statesmen there is invariably to be found, as one goal of the furtherance of prosperity, the demand that everyone be put in the position of participating in the achievements of technological progress, above all through the acquisition of an automobile. In today's industrial society, this demand belongs among the most important criteria of the general standard of living. This is all the more valid since the "house on wheels" is, in very general terms, one of the first steps in the accumulation of wealth for broad social circles.

3. *Social and economic policy,* if, as the citizenry expects, it is to remain honest and consistent in the setting of this goal, must not only make possible the *acquisition* of an automobile for everyone, but also *create all the preconditions for its sensible use.*

4. Next to the *legal regulation* of traffic among these preconditions, a *maximum of technical facilities* for insuring smooth traffic flow and securing property and life for all travelers must be effected.

It is cheap, if not outright hypocritical, constantly to refer to the "slaughter" or "murder on the streets," when one does so from an incomprehensible lack of appreciation for the goal of *making the streets once again into an space of humane mobility.*

5. The preparation of an adequate network of streets in city and countryside demands the *investment of unusually large financial resources.*

The street is a project by the community for the community. However, the motor vehicle driver is aware that he must make a special contribution to road construction. Drivers are already making this contribution annually in the billions—in 1965, 8.5 billion marks.

Policy ruins the chance of making the advantages of motorization maximally available for the good of everyone when it uses major portions of drivers' road construction contributions not for road construction, but for other general state obligations. But not only material losses result from this policy. One cannot use traffic revenues, which climb into the billions, for other purposes without conjuring up mortal dangers for the flow of traffic, that is, without hindering their mitigation considerably. Conversely, it is certain that we face solvable problems, if we make unlimited use of the financial resources inherent in motorization for street construction. Those who desire prosperity for all in the era of motorization need good streets above all. We can have them.

6. The pressure to solve the great common national tasks that face us, which derive essentially from the dynamic of technological progress and the associated prosperity, calls for the *courage and determination to take on many vitally important goals simultaneously in an equal and balanced fashion.* Only when this occurs will it be possible to satisfy the manifold of our common needs on the basis of a sound economic and social order. Here the condition of our streets plays a decisive role.

There exists for the individual person as for the healthy functioning of an entire social organism a large number of unconditional vital necessities. They must be met, not one after the other, but all together and without preference for one or another area. For the modern state, an ordered network of streets belongs among these indispensable foundations of existence.

7. The *traffic chaos* foreseen repeatedly for more than ten years, the result of insufficient measures, has come to pass.

- In our *cities and rural communities* this situation has, through daily traffic jams and traffic disturbances, cost great losses of time, and many victims have lost life and limb as well as incurred daily irritation. This discontent, which all travelers suffer, can only be overcome if the great resources placed at our disposal by the considerable contributions of motor travel are employed without the slightest delay to implement comprehensive ameliorative measures.

- Alongside the critical conditions existing in the cities are equally grave *transport deficiencies in the rural areas.* Overcoming them belongs, in our view, to the established obligations of policy.

- The Federal Republic of Germany is the *transportation crossroads* of Europe. This fact has been taken into account in the *intensified construction of a modern highway network.* In these efforts we must not neglect—as it must equally be one of our political goals—to influence our neighboring countries so that they promote the transportation integration of Europe in the same way.

8. The *simultaneous execution of the most pressing tasks* on these three levels of road transportation is thoroughly possible given the complete utilization of the reve-

nue gained from motor transport *without the imposition of any additional burdens.* The federal parliament and the federal government are thereby called on to develop, in cooperation with the states, a *new, consistent concept for financing road construction,* such that tax policy be administered according to the same principles that are recognized as the basis of social and economic policy.

In all avowals to promote prosperity, the automobile is characterized as an object of daily use, the free utilization of which, as of the radio, washing machine, refrigerator, or sewing machine, should be secured for all. Once, however, the deficiencies in the general budget become an issue, then the use of an automobile is regarded as a luxury and so especially exploited for tax revenue to finance general state obligations. This contradictory orientation of

policy toward the automobile is left over from a transitional period and must finally be overcome by policy itself. Only along this path is the solution to the problems facing us to be found.

9. *Motor vehicle traffic must not,* as a great statesman expressed it, *be allowed to become the stepchild of the nation*—neither morally, in that it is held to be nearly exclusively responsible for the critical state of transportation and for the evil of traffic accidents, nor politically, in that it is forced to bear burdens that are properly borne by all citizens.

The class struggle of social groups and strata, now long since overcome, must not be allowed to reappear in a class struggle among drivers or of nondrivers against drivers. Motor vehicle drivers do not constitute a class unto itself; they represent, because of progressing motorization, the mass of

are not afraid of fire. They are now putting the finishing touches on factories that will fulfill all the preconditions to deliver twenty-four million automobiles, as predicted. . . . The motorization of the populace marches on. . . . Unless all the signs are wrong, the twenty-four million cars really will be built and sold. Not only because enterprises like Ford and General Motors do not invest haphazardly, but also because development once underway cannot be stopped. One cannot withhold the automobile from some, when others already have it. And that is why—according to the simple formula indicated by all relevant market research up to this time—the boom will continue until everyone eligible has an automobile." [1]

The need had been building for decades, and now it could be transformed into effective demand. The attraction of the automotive lifestyle gained added force through the calls for equality in a consumerist democracy: where the same need is attributed to everyone—in this

1. *Auto-Motor-Sport* 18 (1962): 25.

the people and demand no special rights. But for the same reason, they are justified in defending themselves against special obligations, including taxes to the extent that they exceed the obligation of drivers to contribute to street construction. The constitutional principle of equality applies to drivers too.

10. The *automobile* should not be regarded and characterized *only* as a *symbol of prosperity*. On the contrary, one of the most significant mandates of politics is to create the preconditions for the automobile to bring the blessing of technological progress to all, so that progress may fulfill one of its noblest tasks: to pave the way for and intensify friendly relations among persons and peoples.

The current impression made by the automobile stimulates an abundance of pessimistic views concerning its value to humanity.

Motorization is not responsible for this situation, but rather a policy that has so far failed to awaken the positive values and developmental forces dwelling within. Our century has the automobile to thank for its extraordinary economic impetus and the resulting unexpected rise in the general standard of living. The automobile also grants us the possibility to make life nicer, broader, and freer.

It took millennia before humanity was liberated by the motor from the drudgery of the most arduous and slavish toil. The twentieth century is involved in furthering this development for billions of human beings and, in the sense of our great historical development, pushing open the gates to a new age. The automobile is the symbol and tool of this human striving after new goals, which can only be achieved on good roads.

ADAC-Motorwelt, 1965, no. 5.

case, the need to drive—it simply does not do for only a minority to have it satisfied. Once the idea is self-evident that the good life depends on a specific product, then the growth in desire for that product keeps pace with increasing purchasing power—a two-step of cultural transformation and rising incomes. The more people believe that they can improve their emotional and family life and their daily business with an automobile, the more people want to purchase a car, and the more dynamic becomes demand.

Such lifestyle images had sunk deeply into the social consciousness since the classic days of the automobile, and now they penetrated beyond the cities into the villages and beyond "exclusive" circles into the milieu of the buttoned-up, pious petty bourgeois. However various their views otherwise, in the desire for an automotive lifestyle the dairy farmer in Miesbach now agreed with the coal miner in Essen, just as the left-wing student made common cause on the issue with the parson's maid. It is a mass market indeed when people from all walks of life count on an industrial product to give a new sheen to their lives—and this mass market the companies now set about preparing to conquer.

Opel, for one, spared no effort. The company pressed its Kadett on the market not only to win first-time buyers from the Volkswagen plant, but also to prevent those "moving up" from Volkswagen from becoming easy prey to the Ford 12M, which had just come out to attract precisely these customers. The *Auto-Motor-Sport* test report concluded:

> Although Opel cannot afford to outdo (in price) the Volkswagen, let there be no doubt: in the struggle for the remaining reserves of buyers, in the advance toward the twenty-four million cars, the Kadett will offer the VW serious competition. As a challenger, [however,] it will have it just as tough as the new Ford, and it is not yet known to what extent Opel's new model will compete with its own Rekord for new and used car buyers. But there is no going back; the signal has sounded and the great chase is on.[2]

Opel employed the same strategy against Volkswagen that General Motors (Opel's parent company), under Alfred Sloan, had used against Ford in the twenties: arrangement of the products of a single trademark hierarchically. "The core of GM's product policy," Sloan had said, "consists in producing an ascending line of cars graded according to price and quality."[3] This strategy enabled buyers to move up within a single "trademark" model line as they progressed upward in income and personal needs. Meanwhile, marketing experts created for the entire product palette a "trademark personality" (BMW: "the pleasure of driving"; Mercedes: "your guiding star on every street") and, by emphasizing the small differences, garnered recognition—at the expense of the competition—for the product personality in every class. Every firm thus took pains to close the "holes" in its line, to add on above or below, to secure for itself the biggest possible piece of the pie.

Foreign producers, too, had become increasingly aware of this pie, for the internationalization of the domestic market proceeded apace: whereas in 1960 only 9.7 percent of new concessions went to foreign producers, in 1971 the figure was 25.2 percent. All in all, the palette of product offerings expanded considerably in the 1960s, with cars of every shade wooing the customers, so that differences in status, preference, and income now found expression in the differences between

2. Ibid., 28.
3. Quoted in Emma Rothschild, *Paradise Lost: The Decline of the Auto-Industrial Age* (New York: Random House, 1974), 38.

"Believe it or not: It's a VW." Getting away from the notion that VW = Beetle. Advertisement from Der Spiegel, *1965, no. 42.*

automobiles. The producers were equipped to meet the demand, stemming, on the one side, from a rising number of farmers, workers, and lower-level employees who had decided to purchase a car for the first time and, on the other, from a large population of independent, self-employed, and better-paid employees who wanted to move up the scale in automobile ownership.

To move up: this leitmotif sounded in multiple variations throughout the sixties. Educational reform promised upward mobility, increased income expanded purchasing power with every passing year, and all kinds of careers beckoned in offices and plants. What might be called a vertical vision of life was established, namely, the notion that every person's biography leads upward and that those who are attentive to achievement can ascend, step by step, the ladder to prestige and riches.

The hierarchy of automobile models was also an invitation to upward mobility, offering career-conscious individuals a way to move up—to more performance, more speed, more acceleration. Even the Opel Kadett, truly no lively vehicle, attempted to outbid VW in this respect: "There is a lot to it. Even when you stand in front of it for the first time you sense it: this car has something to offer. Forty horsepower makes itself felt." The other firms were not about to let the image of power pass by either, as the shift in advertising rhetoric makes plain: while economy and comfort were emphasized above all else until the sixties, afterward the sporty side of cars took the lead in advertising content. Horsepower figures and acceleration data abounded; cars now leaned into the curves. Common were such slogans as "The first time the light turns green, you'll know what it's got," and "A nose ahead when you need it." Opel's "Only flying is better" became a household saying, Esso put a tiger in the tank, and Ford's best-seller went by the name "Mustang." The power age of the automobile was the 1960s, not only in fantasy but also literally, under the hood: in 1962, only 14 percent of cars were had engines exceeding 1500 cc; in 1973, 49 percent of automobiles were so equipped. All horizons were yet open, progress was speeding ahead, all limitations were to be overcome: that was the spirit of these years, not only in the journey to the moon, but on the street as well.

In 1972, the sixties even caught up with Volkswagen. Wolfsburg had resisted the trend toward product differentiation in performance and acceleration for years, betting all its chips on the Beetle. "There are shapes that cannot be improved," the famous ad boasted in defiance, portraying a VW in the form of a chicken egg. From time to time VW made gestures in the direction of performance—from the VW 1500 to the K70, for instance—and although eventually it did try, with the more powerful "Superbeetle," to catch the rising wave, there was no trademark personality behind the effort. A VW was simply a VW—despite Cinderella's attempt to make a virtue of necessity by ridiculing the appetite for status: "This is a car for people who want to distinguish themselves from people who want to distinguish themselves" (1965). But the decline was no longer to be held off. The spirit of the time, hungry for the future, wanted everything faster and bigger, and the upright fifties were forgotten. Even with more horsepower, brighter colors, and smart blue-jean stripes, VW could not help but recall the "old days"; what, after all, is a Beetle compared to a Mustang? In 1972 Volkswagen capitulated: the Beetle was rolled off onto a siding, and the central station was given over to a new product family, Golf, Passat, and Scirocco. Even the Beetle had to yield to mass motorization's inner drive: not a car for everyone, but for everyone a car.

Traffic Without Space

While earlier it had been chickens, horse-drawn carriages, and potholes that had caused drivers to flush in irritation, now a new disturbance was taking over the streets: other drivers. Motorization was surging, but it came to an abrupt stop in traffic jams; "overfilled streets" and "clogged cities" threatened to make scrap of the newly won freedom. "Traffic Without Space" was the title that Shell Germany, Inc., gave to its first prognosis of future motorization—suggesting the same consequence that the Führer had predicted for the German people in his time: the conquest of deficient space. Now, of course, the front lines were drawn through Germany proper, and the Blitzkrieg was no longer appropriate; it was more a question of skirmishes and trench warfare, to widen the streets to the disadvantage of beer gardens and residences, marshes and maple forests. An internal expansion was on the agenda, governed by the idea of creating penetrable space.

The automobile delivered the power to overcome distances, but not the path. The General German Automobile Club (ADAC) struck upon a fitting comparison in its "Manifesto on Motor Travel" of 1965:

> As it takes both lock and key to compose a perfect technological unity, so do the automobile and the street belong inseparably together. They, too, have to fit each other if motor traffic is to function and all the economic, social, and political advantages contained in motorization are to be secured in their entirety.[4]

The streets, in short, had to be widened until all the cars fit on them; for without a street, a car is like a key without a lock. In the end, an area's penetrability depends on the traffic arteries—streets, rails, canals; the accessibility of destinations, on the network of pathways; traffic volume, on the width of the pathways; and traffic speed, on the impediments in the pathways. Mass motorization demanded mass penetrability; and that is why street construction was the order of the day in West Germany between 1962 and 1978.

Suddenly all kinds of frustrated street planners found themselves trapped in traffic. A chorus of demands swelled, placing traffic policy for the first time in the center of the public interest: More parking places in the city! Away with clunkers on the freeways! Further integration of the provinces! Expressways for through traffic! Millions were prepared to view the world from the perspective of city planners. Expansive regions congealed into mere distances to be covered. The Hofolding Forest, for example, was no longer regarded in its own right as a wooded region where animals live and people go for recreation, or the little town of Wolfratshausen as a residential area where people

---

4. Quoted in Thomas Krämer-Badoni, Herbert Grymer, and Marianne Rodenstein, *Zur sozioökonomische Bedeutung des Automobils* (On the socioeconomic significance of the automobile) (Frankfurt, 1971), 239.

ONE OF THE MOST MODERN ROADS in the world has been given over entirely to traffic in the United States. The so-called "thruway" connects New York to Buffalo. That is a road of the twentieth century. Following the model of the German Autobahn, America is now building its own highway system on a grand scale. The thruway is only one of the many roads in the country for which everything has been considered. Broad, parallel roadways are so arranged that opposing cars cannot blind each other with their lights. Precisely conceived safety features see to it that, despite extreme traffic density and high speeds, the number of accidents is extraordinarily low.

AMERICA IS DOING IT RIGHT. With energy and farsightedness, the Americans are tackling a problem that is causing the gravest concern to responsible men in all the countries of the free world. To meet the great challenges to traffic management posed by the economy and national defense over the coming years, there is only one solution: to correct the deficiencies in street construction by making the necessary resources available. In the United States, President Eisenhower has adopted this cause as his own and announced a program in which nearly 40 billion marks will be spent annually for road construction over the next ten years. The program will allow the road system to keep up with the steady rise in traffic and population in America. General Clay, who organized the famous Berlin Airlift after the war, has been called to chair a commission that will advise the president personally in matters of street construction. It is not true that only dictatorships can build roads. Dynamic and active democracies can do it too.

HOW DOES IT LOOK HERE AT HOME? We are still far from equaling the advanced position of U.S. motorization. Our traffic network is, compared to the American, catastrophically obsolete and backward. That is why Germany has an especially poor traffic safety record. No one has more reason than we to get energetically to work in the coming years, expanding existing streets to handle the traffic and closing the gaps in the highway system.

INSTEAD we are doing particularly little. We are not building enough streets but seek instead to throttle traffic through prohibitions and other bureaucratic means. But the provisions of yesterday cannot address the vital issues of tomorrow.

THE LEVEL OF MOTORIZATION IN GERMANY is strongly linked with the expansion of our economy. Our rising standard of living causes the number of vehicles to grow daily. Throttling will not work. As Henry Ford once said: it is not America's general prosperity that is to thank for all the automobiles; rather, it was the automobile that helped America along its way to prosperity.

GERMANY IS NOT AMERICA. The population of the Federal Republic [at fifty million inhabitants] is only one-third that of the United States. Here it is a question of only 2.3 billion marks, compared to the 40 billion that America wants to spend. But if we, in this as in all other areas of international competition, want to keep up, then this amount requested by the federal minister of transportation is the least that must be done for the maintenance and expansion of our streets.

What do you have to say? Please write to us at the Forum.

Advertisement placed by the Transportation Forum, Frankfurt am Main, in *Der Spiegel*, 1954, no. 49.

live or children play—they were now merely intermediate spaces between points in a superimposed grid, the aim being to get from one to the other as quickly as possible. In other words, the planners made space into a hierarchy of more important and less important areas so as to secure uncongested circulation between the "important" ones. Only through arrogance toward the intermediate spaces could mass penetrability be effected, and it was from this perspective that city and regional planners everywhere got down to business.

The sixties saw the prime of general traffic planning—and no wonder, since in the cities traffic was getting in the way of traffic. "We identify as a typical malady of the times," wrote city planner J. W. Korte in 1959,

> the rush-hour traffic jams in our concentrated urban areas. They cause great losses in valuable time and unacceptable hardships for the entire city economy, and overburden people, vehicles, and streets alike—all of which calls urgently for alleviation. People become nervous and thereby more prone to accident; wear on vehicles increases, as does fuel consumption; and communications as a whole within the compass of the city are increasingly disturbed.[5]

To create penetrability, one had first to sort out the various kinds of street traffic and grant right of way to the automobile: that meant protected areas for pedestrians, discrete paths for bicyclists, separate railways for streetcars, and turning lanes at intersections. The next step was to make main thoroughfares into multilane streets and to synchronize traffic signals on fast streets, which would then be relieved of burdensome cross-traffic through the construction of tunnels and overpasses. To keep the "blood flow" of the city from collapsing on account of "dangerous thrombosis," radial streets were built through the suburbs, connecting the service and goods center of the city with the employees and consumers on the outskirts, and tangential streets were constructed through the areas bordering on the inner city, to distribute incoming traffic over a larger area.

The bureaucratic language of the national administration presented this dictatorship of penetrability as completely harmless: "The states demand that expansion programs for city streets be drawn up with the goal of increasing as much as possible the surfaces devoted to traffic (including surfaces for parking), as well as disentangling the various

5. J. W. Korte, ed., *Stadtverkehr—gestern, heute und morgen* (City traffic—yesterday, today, and tomorrow) (Berlin, Göttingen, Heidelberg, 1959), 9.

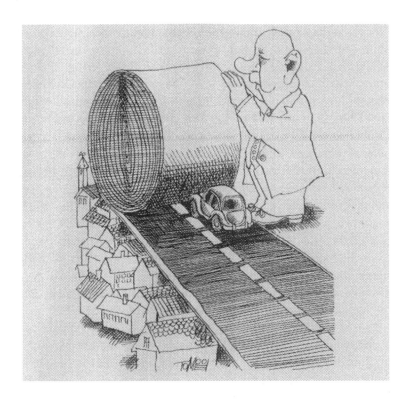

types of traffic and traffic flow by means of additional traffic levels."[6] If the city made for the automobile was not yet a reality, the idea had taken hold by the end of the decade. Indeed, the city made for transport did telescope the intermediate distances, allowing commuters to circulate between work and home; subway tunnels were also being dug, and streetcar lines snaked throughout the city. The more accessible the space, the better the city planning. Urban space deteriorated into a distance to be overcome.

Thus were city planning and national highway construction rung in during the sixties. For the countryside was being emptied precisely to the extent that the concentrated areas were growing denser. The rural problem was not too much traffic, but too little; not disentanglement, but inclusion. With the forcing of production and consumption in the

6. *Die kommunalen Verkehrsprobleme in der Bundesrepublik Deutschland. Ein Sachverständigenbericht und die Stellungnahme der Bundesregierung* (Regional traffic problems in the Federal Republic of Germany: An expert report and the position of the federal government) (Essen, 1965), iii.

hothouse centers to ever new growth, the differential between city and countryside had become more pronounced. In line with the goals of the Regional Planning Law of 1965 to create equal living conditions, national roads and highways were to link population centers and rural areas, thus tying the countryside into the currents of an industrialized way of life. (It is for this reason—to connect rural life to the urban pulse—that the first great long-distance streets of modern times, the *routes royales,* were built in France before the revolution.) Labor power would flow to the cities and investment capital to the countryside, and on the weekends there would be tourists: in any case, streets were to be extended, hills leveled or bored through, valleys filled in—and the sweeping curve of an overpass always suited the landscape better than a flat intersection. No village would be too distant, no citizen too cut off; in the famous words of Traffic Minister Georg Leber, a highway onramp would never be more than twenty-five kilometers away. The more penetrable the space, the better the rural planning. The homeland had deteriorated into a distance to be overcome.

But space is most penetrable . . . in the air. Why remain on the ground, when one day it will be possible to travel easily through the air? Futurists were already farther along in the sixties, depicting a whole new space for traffic:

> The moving walkway conveys us two hundred meters from the front door to a bus stop; the electric bus, a kilometer and a half to the nearest helipad; the helicopter, finally, with a scheduled intermediary stop, to the airport, thirty-five kilometers from home. The entire trip takes no more than twenty to twenty-five minutes—a distinct possibility for many of our larger cities within a decade.[7]

Such were the projections in 1972.

7. Robert Brenner, *So leben wir morgen. Der Roman unserer Zukunft* (Our life tomorrow: The novel of our future) (Munich, Gütersloh, Vienna, 1972), 95.

Desires

"A new symbol of classical beauty: Adler Standard." Advertisement for the International Automobile Show, Berlin 1928.

Independent as a Lord

Searching for a fitting image for the meaning of the automobile, the French philosopher Roland Barthes came upon one of those comparisons that illumine the nature of a thing: he called the car "the Gothic cathedral of modern times." This is a surprising metaphor on first glance, for what do automobiles have in common with these light-filled edifices yearning toward the heavens? But a second glance uncovers the point: automobiles, like cathedrals, are symbols of a culture. However acute in vision, one would be blind who stood before a cathedral and saw nothing more than a shelter from wind and weather for the faithful. And one would be just as blind who stood before a shimmering automobile with its engine humming and saw only a vehicle for transporting people and their goods. As the cathedral is not merely a shelter, so the automobile is more than a means of transport; automobiles are, indeed, the material representation of a culture. Although both creations contain considerable engineering artistry, under the technical design lies a cultural plan in which the assumptions of an epoch find expression. The engineers, with their calculations and drawing boards, create something that is important to the public and toward which the energies of an epoch are directed, whether that be the love of God or the love of speed.

Far from being a mere means of transport, automobiles crystallize life plans and world images, needs and hopes, which in turn stamp the technical contrivance with a cultural meaning. In this interchange, culture and technology prove mutually reinforcing. Technology does not simply fall from the sky; rather, the aspirations of a society (or a class) combine with technical possibility to inject a bit of culture into the design like a genetic code. Yet neither do lifestyle and desires emerge from the thin air of culture; instead they coalesce around a given technology. A technological invention is often accompanied by cultural creativity. The invention gives dominant motifs a new material form, whereupon new motifs are invited to develop. What ideals, what life projections have accumulated around the automobile? What drives have marked its technical design, and what new lessons has it revealed? For the lessons surrounding the automobile are still new, and are not, as the occasional triumphal automobile book announces on page one, to be traced back to the ancient Greeks or anywhere else. The history of technical development proceeds hand in hand with the history of cultural acquisition. Although the latter offers technology a place in life, it can also turn and make the technological product obsolete when the feel of life in a new epoch is no longer reflected in the technology. Technical achievements come and go; what was once learned from them can be forgotten. The Gothic cathedrals, though, unlike the automobile, built for eternity, became antiquated in the fifteenth century; yet they still perform masterfully in the present, helping tourists and schoolchildren decipher the spirit of a bygone age.

Masters over Schedules and Routes

For those in the nineteenth century who were particular, traveling by railway imposed unwelcome demands. While some salon cars certainly resembled a grand hotel on wheels, the luxury of first class could not conceal that gentlemen from more refined circles in fact forfeited sovereignty: they were inmates in a system of mass transit. Because of technological progress, they were forced to give up the carriage, with its potential for improvisation and freedom, and subordinate themselves to external constraints. "Traveling by railway," came the warning from *Meyers Konversationslexikon* of 1850,

> demands the most punctual arrival of travelers at the train station, because the steam engines wait for no one; and canceled tickets and baggage claim checks must be carefully preserved. Moreover, one must take care to remain on the train at stations where it is not stopping for

at least five or ten minutes, because the train can easily roll away, leaving the traveler in the lurch.

Racing furiously for the train; lost in a cloud of smoke and noise and desperately seeking directions to the correct car; fumbling the ticket out of a jacket pocket for the surly conductor; then to be shot rattling across the country—and all this under the penetrating gaze of the common people. No, this was an affront to proper order, a blow struck against the good old days. The railway delivered the traveler up to an anonymous machine and was degrading to a cultivated style of life. With the railway the independent traveler became a dependent passenger, conducted from place to place like a piece of freight. Passengers were, indeed, "transported," a word that until now had been applied only to prisoners or a salesman's wares. To be a mass transit passenger, an indifferent appendage equal to everyone else in the enormous wheelwork of the railway, ran against the grain for those refined individuals who, relying on their own property, were accustomed to indulging an independent and stylish enjoyment of life. "The railway," Otto Julius Bierbaum complained,

> transports us—and that is the direct opposite of traveling. We are condemned to passivity—whereas traveling signifies the freest activity. Traveling is throwing off the yoke of rules. The railway binds us to a timetable, makes of us prisoners of regulations, locks us in a cage, which we are not allowed even to open, let alone leave, when we please. Between telegraph wires—symbols of this entangling of our personal freedom—we are hauled at a speed that completely eliminates the possibility of welcome sights, not from one place, but from one train station, to the other. . . . Whoever calls that activity "traveling" might just as justifiably pass off a parade march as a stroll. The whole purpose and benefit of letting oneself be hauled in this way is based in overcoming distances. . . . The old drinking song "Stupor, stupor, you're my pleasure, stupor, stupor, you're my desire," would be a worthy hymn for the undertaking.[1]

But now the railway had a competitor: the automobile, which appeared on the stage just as the refined world was coming to terms with travel by train. The memory was still alive, however, of the time of carriages, when one was one's own master and could drive a private coach with pride, and this memory deeply colored the attitude toward the new motorcars: it seemed the glorious days of carriage travel had

1. Bierbaum, *Eine empfindsame Reise*, 269.

Faces of the railway. The rush for cars when the excursion train departs Paris for the sea, 5 francs. Caricature by H. Daumier, 1852.

come again. Automobiles promised to resurrect the old independence of self-propelled vehicles, to help individual authority regain its own, for they offered emancipation from the inconveniences of the railway: the regimentation of the timetable, the compulsion of the unwavering rails, and—not least—the perspiration of the crowd. The transition from carriage to railway had jeopardized the confidence of status-conscious groups. Now the sovereignty lost could be recreated by moving from the railway to the automobile; indeed, enjoyment would be even fuller on this new, mechanized plane. No more being ordered around by shrill whistles; no more surrendering the baggage into who knows what kind of hands. Gone forever the undignified existence of a passenger!

> A blissful prospect: never to be plagued by the fear of missing the train. We will never have to cry for stewards, never have to count again and again, one, two, three, four—did he bring everything? My God, the hatbox! Is the umbrella there too? We will never run the risk of being locked in a compartment with insufferable people, where the windows cannot be opened even in oppressive heat if someone is along who suffers from a fear of traveling in trains.

Bierbaum indulged his obituary for the railway as he headed off with his wife and, of course, a chauffeur for Italy. Gazing at his Adler-Phaeton, his luggage secured to the back, he reminisced about his time as a "box person" in a train compartment. His newly arrived automobile opened up entirely new vistas to him:

"You are looking for your suitcase, good sir? It's over there somewhere—and the lady's hatbox too . . ." Caricature by H. Daumier, 1843.

We will decide ourselves whether we drive fast or slow, where we stop, where we want to pass through without delay. We will be in the brisk, fresh air for days on end. We will not drive in dark, terrible caves through the mountains, but over the mountains. In short, gentlemen, we will truly travel, rather than have ourselves transported.[2]

It appeared that with the automobile had arrived nothing less than the end of the age of transport—"transport" being the essential characteristic of the railway, because it moved masses of people from one place to another, organized according to the unyielding logic of a centrally directed apparatus into multiplicitous, daily-recurring movement: locomotives, tracks, and schedules. Those who used the railway, playing their part in the progress toward a greater mastery of space, had to give themselves over for good or ill to the logic of this apparatus, this "iron cage of subjection" as Max Weber might have called it. These objective constraints ran counter to the love of individuality professed by elevated sorts, who placed great value in shaping their lives by their own decisions and developing their own unique patterns of expression and behavior. This was the sentiment of the fashion writer Countess Ida Hahn-Hahn, for example, who in 1841 tried to see the railway as a "travel outing"; but, she determined,

for a genuine trip, I find it altogether unseemly for a person. Through train travel one is degraded into a bundle of goods, and forfeits the

2. Ibid., 20.

On the passenger train. Between a butcher and a drunk. Caricature by H. Daumier, 1839.

individual senses and independence. Normal human consideration is not extended to a bundle of goods; the only obligation is to send it on. . . . The steam locomotive aims only at leveling and centralizing—the two obsessions of those who call themselves liberal. . . . All limits, sense of place, pleasures, and needs are likewise leveled. For a pittance, old and young, fine and common, rich and poor, human and beast alike glide along behind a steam engine.[3]

The railway confronted bourgeois-aristocratic circles with a dilemma that runs throughout the history of industrialization: while the increasing mechanization of social life did indeed open up unexpected possibilities, it nevertheless threatened that well-tended subjectivity, that self-consciously personal lifestyle that, particularly since the romantic era, the bourgeoisie had developed to shield themselves from mechanization. On the one hand, the locomotive fascinated: mechanical force made it vastly superior to the horse, limited by its corporeality and therefore subject to exhaustion; and the pounding uniformity of its movement inaugurated a new mastery over more extensive space. On the other hand, the railway inspired discontent, dismay, and melancholy, because a bit of the art of living fell beneath its wheels, it proved so able to shrink the distance on which cultivated superiority was based—the distance not only from the common people, but also from a life subject to regulation.

The automobile, then, presented the possibility of escape, for it re-

3. Quoted in Manfred Riedel, "Vom Biedermeier zum Maschinenzeitalter. Zur Kulturgeschichte der ersten Eisenbahnen in Deutschland" (From early Victorian times to the machine age: On the cultural history of the first railroads in Germany), *Archiv für Kulturgeschichte* 43 (1961): 119.

turned resonance to a world of sentiment that in the age of the railway had become obsolete: the attraction of travel guided by nothing but individual pleasure and mood. By uniting mechanical force with self-determined mobility it opened the way to harvesting the fruits of the transportation revolution without having to pay the price. It was almost as if a new outbreak of individuality loomed:

> We want truly to travel again, as free gentlemen, with free self-determination, in the freedom of the open air. And the fact that we will have more to do, that we will be faced at every moment with the need to make decisions, is an advantage of this resurrected art of travel. Traveling in an automobile includes not only a massage for the body, but one for the spirit as well, and that is precisely why . . . the power of invention, the perception, the processing, the internalization of external impressions, is intensified.[4]

A genuine auto-mobile, a self-propelling craft, which troubled one with neither schedules nor preset routes, a pledge of freedom and individual pleasure—that was the perception that made the automobile so attractive against the background of train stations and tracks. Thus did the excitement over a new technology grow out of disappointment with the old, thus did yesterday's critique of progress prepare the way for progress tomorrow. The feeling of independence was born of the contrast with the railway.

The Individual Is King

In fact, according to a 1974 speech by the president of the German Automotive Industry Association, J. H. Brunn, "the desire to be master of time and space without dependence on schedules was not invented in an automobile factory. It accords with the nature of the modern person and comes from the consumer. Everyone should be able to use the means of transit that best suits his or her individual needs."[5] As carefree as can be, he challenged critics of the car (this was just after the first energy crisis) in the very title of his speech: "The Automobile Is Another Bit of Freedom." Although "the nature of the modern person" invoked by the chief automobile manufacturer was fully unknown to our grandfathers, he was addressing the sentiment that underlay the

4. Bierbaum, *Eine empfindsame Reise*, 292f.
5. J. H. von Brunn, "Das Auto ist ein Stück mehr Freiheit" (The auto is another piece of freedom), Speech to the VDA-Mitgliederversammlung, Baden-Baden, September 27, 1974.

automotive economy: the mass desire to be free of fixed rails, schedules, and other people.

Again and again traffic analysts have plunged into intricate investigations of why so many people prefer a car to the bus, streetcar, or subway, only to arrive at the obvious: in order to be independent in time and space and also socially, by being able to choose one's own means of transport. Bierbaum's excitement in his day over the self-propelled vehicle has in the meantime become a mass sentiment; people today forsake the streetcar for gasoline carriages in droves. Indeed, the automobile is all the more enticing now because, in contrast to Bierbaum's times, the question has long since ceased to be one of replacing the occasional train trip with the pleasure of an unfettered drive in an automobile; the issue now involves preserving some measure of independent mobility inside the machinery of daily transit. As the "forwarding of people like bundles of goods" became a virtually inescapable fact of daily life, the automobile—if no longer in the actual memory of a cultivated carriage journey, then at least in the hope for a bit of freedom in the general business of transportation—came to be the focus of desires for individualization.

With generalized mass transit grew an urgent need for the automobile. Countless passengers preferred, like Bierbaum of old, to drive themselves. Why stand in the cold at a bus stop, why tediously patch a route together from tram line to tram line? This sentiment nourished the call for "freedom of choice in transit," which has drowned out the voices of other, divergent interests in transit policy for decades, causing mass transit to lose the competition with the automobile. In response to new pressures for mobility, automobile production increased and, contrary to every intention, stimulated a new round of clamoring for automobiles. Consumer expectations of the sort aroused by advertisements like that for a 1955 jalopy, the Maico 500, caused the demand for cars to soar: "The daily irritation—commuter trains filled to overflowing, overloaded streetcars, oppressive crowds—no wonder many people arrive at work in a bad mood. Is there no remedy? Of course there's one: motorize yourself! With Maico, for they are the cars of tomorrow."

This desire was not pulled out of the thin air of cultural significations, but was based on realities already incorporated in technological products. Cultural significations cannot make their mark arbitrarily; they must first exist as a possibility in technological form and function. The characteristics of the automobile confirm the idea of independence and allow it to appear as natural. Unlike the railway, an automobile can be acquired privately and is therefore always at the owner's disposal. Because it is not bound to the tracks (and sometimes not even dependent on roads), it can be driven almost anywhere. Because one person steers it, it need not be shared with others; similarly, it can satisfy any particular desire of the driver in terms of speed. Finally, there is the multifaceted character of modern cars: they are just as good for displaying status as for being loaded with surfboards. Equipped with this dowry of design, the automobile is fit for marriage with the desire for an unfettered lifestyle, a desire that can be undermined only by the experience of one person's craving for freedom colliding with that of another, resulting in traffic jams everywhere.

It is not at all accidental that the automobile is engineered to service individual needs. A design gradient favoring individualization runs through the whole history of modern technology. From major machinery to household appliances, this tendency is always evident; the transition from the railway to the automobile may be the most spectacular example, but only because the shift from churchtower clock to wristwatch, cinema to television, community laundry to washing machine, or adding machine to pocket calculator was less obvious (though the transition from mainframe computers to personal computers is not likely to be outdone in any respect). The industrialization of human activity—of mobility, time, visual entertainment, washing, calculating, and communicating—seems to establish itself first of all in large settings requiring collective use. Only when such innovations acquire a more compact form do they become familiar, rather like a comfortable pair of slippers, and begin to leave their mark on daily appearances and gestures.

The model of transformation seen in the automobile is everywhere: what once was a public commodity is now assimilated to private ownership; what once had to be sought out at a particular place now becomes everywhere available; and what otherwise ran according to an impersonal time plan can now be had anytime. We find in the production of implements that can be used by one person, anywhere, and anytime—that is, in a manner that is independent socially, spatially, and temporally—a deep-seated purposiveness of technological development. Consider, for example, the progression from cinemas to the console television to the portable television to the video cassette recorder: what once could be seen only collectively, in a distant hall and at a predetermined hour, can now be admired by everyone on his or her own, in the camper or in the living room, morning or night.

It is given in the structure of many advanced appliances that they in no way interfere with the private caprice of the individual. Agreement with other people is rendered superfluous, dependence on a particular place eliminated, and the individual can ignore all impersonal temporal rhythms—therein lies the advance. The ideal buyer of these appliances, whom it is their purpose to serve, is the individual as king, the sovereign who wants to use them according to personal discretion and without social, spatial, or temporal constraints. And such an atomized collection of customers is ideal for the industry; a mass market is attainable only through individualization of the appliances to be sold.

Technological development—embodied in screws and switches, tools and, ultimately, appliances—has reached a state which, although self-evident to modern culture, nevertheless signified a profound transformation of the traditional conception of life and has only gradually over the last two centuries become reality: namely, one in which the individual is preeminently and completely grounded in his or her own right, and all ties to society and to nature are subordinant to one's private decisions. In this scheme, the individual is not part of a whole—whether of a tribe, as for a Sioux; an ancestral line, as for a Kikuyu; a hierarchical social system, as for a Hindu; or a social order ordained by God, as for the European absolutist—or even of a village, family, or household. No, each person is master of him- or herself. Without getting involved in the details of Western intellectual history, we can note that the word *individual* has been common in its current meaning only since the end of the eighteenth century; the former meaning, of course, is found in the sense of "what an individual," suggesting the skepticism once reserved for persons lacking any social obligations. This basic category of our culture also shapes technological development, for in tools and machines what we consider to be fundamental finds expression. Technology is the material reproduction of a culture.

But not only that. In the progression from major system to household appliance, technological development, programmed by the cultural code, strives to make this independent individual materially possible. Automobile or washing machine, video recorder or microcomputer—all are the realization, in the concrete form of an apparatus, of a cultural ideal that has long since migrated from the world of philosophical thought into people's emotional world. Technology fulfills the desire to leave behind burdensome social, spatial, or temporal ties and become one's own master.

The idea of "another bit of freedom through driving," to use the slogan of the automobile industry association's president, finds resonance as long as large numbers of people are steered by the modern ideal. But the more it achieves reality through mass motorization, the more unavoidably the desires for independence slowly change their hue: rubbing one's tired eyes, one discovers a new form of dependence behind the independence gained. Ultimately, all of these "independence machines" depend on streets and power lines, pipelines and radio waves, which in turn bind the individual with multiple ties to industries, power plants, drilling rigs, and broadcast stations. Supply networks and production apparatuses must be called into being to supply us with another increment of freedom in our private lives—a dependent independence, however paradoxical that may sound. While Bierbaum, glad not to be transported in a train compartment like freight, toasted the automobile with hymns of freedom for having released him from existence as a passenger, it dawns on us, driving eighty years later in tight columns on the city freeways, that even self-propelling vehicles can form a transit system organized according to impersonal requirements. With Bierbaum we became drivers rather than passengers, but since then we have metamorphosed once again from drivers into passengers, even if self-propelled.

Riding the Iron Steed

It was not, however, the automobile alone that gave the joy of independent mobility a place in society's affections. Another invention that appeared almost simultaneously shared this honor—namely, the bicycle. In the same years that Carl Benz was testing his gasoline-powered carriage in Mannheim, John Kemp Starley was getting the bugs out of the low-wheeled cycle, an invention that was to ring in mass distribution of bicycles. One had, indeed, dared to ride the high-wheeled velocipede only at the risk of one's neck; balance was always precarious, since the center of gravity was in the front and crept ever higher the bigger—and therefore faster—the front wheels became. Because a slender tree root was all it took to throw the rider head over heels into the dust, at best only athletic gentlemen attempted them, in riding pants and with a well-upholstered helmet on their heads.

In Starley's design, the driver's legs were suspended a comfortable distance from the ground between two medium-sized wheels, and the chain drive allowed even unathletic riders, soon on air-filled tires, to glide smoothly down the street. A real bicycle craze broke out only as

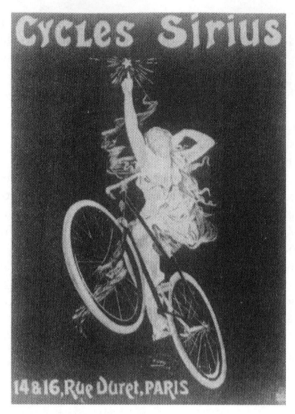

Poster by H. Gray, 1899.

the turn of the century neared, with young and old alike swinging into the saddle. The police had their hands full maintaining a surly authority over the fleet-wheeled people. After the American import trade brought widespread price reductions, the bicycle also became accessible for less well off groups: of the approximately one million bicycles in use in 1903, some 30 percent belonged to workers.

Riding on their iron steeds, many now enjoyed a mobility formerly reserved to the fancy gentleman on his high horse. One's own forces were multiplied in an ingenious interaction between muscle power and mechanics; traveling smartly along, one left the pedestrians behind, elated by a newly won power over space. Zola depicted it in his novel *Paris:*

> The two let their machines carry them down the hill. And then this happy rush of speed overtook them, the dizzying sense of balance in

the lightninglike, breathtaking descent on wheels, while the gray path flew beneath their feet and the trees whisked past at either side like the slats of a fan as it unfolds. . . . That is the endless hope, the liberation from the all too oppressive fetters, across space. And no exaltation is better; hearts leap under the open sky.[6]

The speed is intoxicating, the mobility liberating, the exertion inspiring—that is the experience of riding a bicycle, and it was now becoming common throughout society. It is hardly surprising that the bicycle also drew its attractiveness from a contrast with the railway.

The bicycle is subservient to no time schedule; it is free. It does not follow the beaten path, rather roves along a thousand freely chosen paths. At every hour, in every direction it carries its rider. It serves nothing but individual need; it does justice to the endless variety of human desires and endeavors.[7]

These were the same feelings as the automobile inspired: here too was the pleasure in unfettered mobility, but with an incomparably more modest vehicle and therefore accessible to much of the population.

The bicycle offered liberation from the regimentation not only of the local train, but also of daily life. Whoever seeks to distance himself can have two reasons for doing so: to get away or to arrive. To get away to the distant metropolis, to flee the oppressive conditions and lack of prospects at home—this desire to change one's lot through mobility had already been awakened by the railroad. How many village strolls led to the train tracks, how completely the train station replaced the village oak as a meeting place! It was the magic of redemptive distance that enticed. To escape the watchful eyes and stubborn rules of daily life, even if for only a couple of hours—this desire could be fulfilled with a bicycle. The young could escape the nagging of their parents and the workers the oppression of their cramped apartments; the bicycle delivered the sheltered daughter from her knitting and the pale clerk from the company books.

A forceful step on the pedals also saw the birth of a new self-confidence, and with the lively bicycle came often enough a lively spirit of enterprise. Since it had suddenly become so easy to get away and hold

6. (Paris, 1898), 389; quoted in Hans-Erhard Lessing, ed., *Fahrradkultur* (Bicycle culture), vol. 1: *Der Höhepunkt um 1900* (The climax around 1900) (Reinbeck, 1982), 5.
7. L. Bertz, quoted in ibid., 6.

one's own through distance, many felt their inner independence fortified, thus gaining a previously unknown sense of freedom. The new movements required—the pedaling and balancing, not to mention the very fact of bodily exertion in public—also symbolized a kind of moral liberation, especially for women: "What must absolutely go in the junk room, for starters, is the corset. Deep, brisk breathing, as riding a bicycle requires, can happen only given full expansion of the breast. How can the poor breast expand if it's stuck in a plate of armor? . . . The best and freest feeling," the lady cyclist boldly continued, "comes without reservation from a completely unconstrained upper body. For me, it even makes a great difference . . . whether I ride completely unbridled or with—even if it's very loose—a brassiere."[8] The bicycle became a symbol for the "new woman," who had shed her fetters and started on her sovereign way through the manifold offerings of the world. This escape from oppression provided a model of personal experience that the automobile could tap only decades later, when it became accessible to people who believed it allowed them to flee all manner of oppressive conditions.

While excitement was building over the automobile in terms of distant trips, bicyclists were taking pleasure in the new accessibility of nearer goals. The bicycle enlarged the immediate vicinity and multiplied the destinations that could be reached in a short period of time; whether to the factory to work or the lake for a swim, to church in the neighboring village or a flirtation in the woods, in the bicycle saddle one felt oneself the master of one's native territory. For the first time an achievement in transportation technology invigorated local life—for the railway had sooner enticed travelers to regard themselves as masters of the nation (and later, the airplane as masters of the world). Rural roads, moreover, increasingly empty because of the draw of the railway, were now becoming more lively; where traffic had dwindled to the occasional rural wanderer, now travelers were once again on the move from place to place. Perhaps the railway had opened up space on the level of the nation, but the bicycle was opening space at the local level.

Contemporaries were not blind to the fact that this feeling of freedom cost much in the way of panting and sweat. Indeed, because it required physical exertion, the bicycle was unsuited as a symbol of class; a privileged status, after all, meant to have impersonal energies

8. Ibid., 20.

at one's command, with others doing all the sweating. It was therefore with a note of defiance that Wilhelm Wolf remarked in his 1890 book *Fahrrad und Radfahrer* (Bicycle and bicyclist):

> Whoever is in the happy position of having a horse to ride or a horse and carriage commonly looks down on the bicyclist with a tinge of compassion; it seems to him nobler to be conveyed by animal power than to accomplish movement through his own exertion. All that "kicking about" does not appeal to him, for he cannot help but think that the bicyclist, in pedaling his machine, is executing essentially the same leg movements that our dear God prescribed everyone for walking.[9]

If for no other reason, the bicycle was not quite presentable in the refined world. Thus it failed to rise to the level of a symbol of social superiority, which would have inspired a mass desire to imitate.

The defect of physicality worked even more strongly to the bicycle's disadvantage as the use of motor power became widespread. Because the bicycle did not move without muscle power, it remained an outsider among the new technologies with their aura of progress, for it was precisely in the substitution of mechanical power for muscular exertion that the point of progress was recognized at the time. Eugen Diesel, in his description of the three-month waiting period for his first automobile, stated flat-out that

> bicycling was a wretched substitute in the interim. I did indeed go faster than before; I felt myself to be an automobile—I imitated the shifting of gears, acceleration, adjusting of the ignition, and nearly succumbed to the hallucination that I was a driver. But it was a damned disadvantage that the motor was missing and the pedals would not go by themselves.[10]

Sitting in the bicycle saddle, his legs kicking, the young Diesel imagined himself an automobile driver: shifting, pressing on the gas pedal, letting the engine roar. How many later generations have not experienced the bicycle exactly like this!

There was nothing to be done about it, though—the bicycle fell victim to the contemporary view of what constituted technological progress: overcoming physical limitations through the power of the motor. With no regard for the fact that technological progress was contained

9. Wilhelm Wolf, *Fahrrad und Radfahrer* (Bicycle and bicyclist) (Leipzig, 1890; reprinted 1979), 5.
10. Diesel, *Autoreise 1905*, 19.

even in parts like the bicycle chain, most people regarded motors as the very essence of progress; they contained the magical promise of putting a yoke on the apparently inexhaustible energy of nature and doing the work for human beings. Even though the bicycle multiplied bodily energy extremely efficiently and broadened people's arena of activity many times over, it remained captive to the defect of corporeality; over the long run, therefore, it would be at best a disagreeable substitute vehicle for the nonmotorized.

In retrospect, though, it must be noted that not only did automotive technology—with the chain, the hub, and the air-filled tire—reap the gains of bicycle technology, but the attractiveness of the automobile in particular was nourished on the feeling the bicycle inspired. The iron steed extended the boundaries of spatial experience for the masses and so stimulated desires for increased independence of movement. Mobility in the immediate vicinity was now a fact of life. It was from this reservoir of popular experience that motor vehicles drew much of their attractive power; they promised, after all, an unbelievable increase in mobility. The bicycle, then, in more ways than one, was simply the first along a path that, followed by the motorcycle, ended in the small car just as soon as the purse allowed. The bicycle mobilized desire for an automobile.

Little Escapes

As time went on, the joy of fast-moving freedom, established in daily practice and desires by the bicycle, came to color attitudes toward the motorcycle and automobile as well. The occasions on which people felt a need for wheels to get away for a while, the opportunities self-propelled vehicles have offered for some measure of liberation from oppressive circumstances, are countless. The husband breathes a sigh of relief as he sinks in behind the wheel, having slammed the door on his wife's biting words; overjoyed to have escaped the workaday grind once more, the young worker throws his machine in gear on Friday evening and races off into the open potential of the weekend; relieved, the young mathematics student, academic drudgery causing his tiny garret to close in about his ears, revs his Citroën Deux-Chevaux before making his way to his favorite bar. A change of scene, the chance to breathe different air for a while—such are the motivations, under the leitmotif "flight from obligations," behind the automobile's summons.

The same motif tumbles abundantly from the screen and the pages

"500 Marks—and Whitsuntide in an Opel." The
automobile as a symbol for unfettered joy in life.
Advertisement from Berliner Illustrierte Zeitung,
1929, no. 38.

of literature: the flight into vacation from the routine; the flight of the
young from parental proscriptions; the flight of the criminal from a
closing trap. In film and television the appearance of a car signals a
change of scene; the hero screeching away in a car has replaced the
cowboy of old riding off into the setting sun, majestic music rising in
the background. Especially in American literature of the fifties and sixties, the automobile symbolized the feverish freedom of breaking loose
and getting away. In what is arguably the most famous example, Jack
Kerouac's *On the Road,* the hero of the "beat generation," Dean
Moriarty, completely absorbs the stimulation of the highway's promise

of infinity. And in John Updike's *Rabbit, Run,* Harry Angstrom sits cramped behind the wheel, tormented by the fear that his life, with the whining children at home and his wearisome job selling used cars, has reached a dead end in his disgruntled little town. He takes off for the virginal, easygoing life of the south: "He wants to go south, down, down the map into orange groves and smoking rivers and barefoot women. It seems simple enough, drive all night through the dawn through the morning through the noon park on a beach take off your shoes and fall asleep by the Gulf of Mexico." [11]

Car commercials, too, use this reservoir of significations, indeed, do their best to keep it alive, because it wraps the automobile, as a commodity, in a cloak of meaning that, by stimulating buyers' desires, opens up their purses. The Toyota Land Cruiser leaves boulders and streams easily behind and then masters a wild rapids: "Toyota gives you the freedom to get off the beaten path!" The commercial stresses that a Toyota is not simply a means of transportation, but stands for a lifestyle. To the car are attributed nonmaterial characteristics that derive from an obviously inexhaustible world of locales, where the conqueror of impassable terrains, the natural man far removed from civilization, the adventurer facing endless obstacles, appears as the hero of freedom and independence. The commercial, by integrating the product into a system of established significations, that of the adventure saga, itself becomes clothed in those same significations. Hence a single commercial image, in moving beyond the product's technical usefulness, illustrates the logic of advertisement: the world of products is translated into a world of significations, and vice versa, and all possible significations appear to be purchasable in the form of products.

Just by clothing the product in the qualities of freedom and independence, moreover, the ad conveys information about the character of the Toyota owner: a man "who wants to get off the beaten path." The ad copy even conflates the two, the car and the owner: the car is so rugged that it delivers independence from "mud, snow, and inclines," and, accordingly, its owner proves himself capable of overcoming the conventional rules and routines of daily life. The character of the Toyota mirrors that of its owner: both love freedom and neither is conformist. It makes no difference that the car will probably never be driven along the Amazon; its symbolic power works amid the asphalt and traffic lights of the city too.

11. John Updike, *Rabbit, Run* (New York: Alfred A. Knopf, 1970), 25.

It is obvious that commercials do not draw their power of persuasion just from the inventiveness of graphic artists; rather, they sound the very melody that originated with the bicycle and then accompanied the use of automobiles and motorcycles: the joy of minor emancipations thanks to easy mobility. In this realm of experience, freedom is abbreviated to freedom of variation, change comes to mean breaking away, and solutions are to be found in leaving everything behind. To have no wheels—the thought is coupled with fears of remaining stuck in the given, of being subject to the controlling gaze of others, of losing oneself in the daily grind. The desire for change and emancipation is objectified in the automobile. It is no wonder that for many young people the acquisition of a car is a prerequisite to being accepted in the world of adults, or that women may, on occasion, assert their freshly won self-confidence with a sports car.

Mobility and distance can be liberating; this experience, however, stems from two contrasts that lend these freedoms their particular power of attraction. On the one hand, as the early history of the bicycle reveals, the fascination with moving forward rests on the counter-experience of immobility, the unavoidable attachment to static conditions of life; on the other hand, the pleasure in distance relies on the fantasy that out there at an appropriate distance a completely different life beckons. Would not these preconditions grow increasingly misleading as mobility becomes a social norm that allows everyone to be constantly in transit? If everyone is breaking free, the joy of distance may well decline; if all are driving off in search of foreign experience, it is natural that those very distinctions for which they quest become hazy, and distant goals prove strikingly similar to home.

Universal mobility takes the magic out of distance. It is not by accident that Toyota has to reach into the extreme distance—the jungle—for its commercial; where else are the paths not already beaten? Nor is it happenchance that Updike's Harry Angstrom, having left his provincial nest behind for the Gulf of Mexico, loses his way in a nighttime chaos of signs, motels, and highway interchanges, so that, driven crazy but sobered, he turns back and, nodding off from exhaustion, finally mistakes the automobile itself for his lost destination: "He thinks again of his goal, lying down at dawn in sand by the Gulf of Mexico, and it seems in a way that the gritty seat of his car *is* that sand, and the rustling of the waking town the rustling of the sea." [12]

12. Ibid., 40.

Victorious Speed

And everything swarming with people, with men in blue peaked caps
and women in flowing veils, their eyes alight as if from fever and always looking at a gigantic blackboard, or at a black or blue or white
something that, one, two, three, comes and goes with a thundering din
like the devil himself . . . ; one automobile chases after another, often
with two struggling side-by-side, roaring past the people. Farther, just
farther. And once again new numbers appear on the blackboard; the
public calculates. It calculates who in this titanic battle of horseless
machines made the best time, who was fastest. . . . The battle has been
rolling on for hours already, kilometer after kilometer, lap after lap. A
bitter contest. Already gaping blanks are to be seen on the board, for
the least fortunate are gradually disappearing from sight, now lying
along the way with broken machines or scuttled forevermore into the
roadside pits. . . . There a black point is approaching like a cloud,
growing larger and larger, and the point becomes a rattling monster,
lost in the howls of a thousand voices and surrounded by a thousand
people shaking, in the uproar, the victor's hand.[1]

An electrified public, speed, farther and farther, a bitter contest between catastrophe and glory, one against all and all against one, and
then—the triumphant victor, swallowed up in a frenzy of jubilation

1. *Berliner Tageblatt,* January 4, 1909.

and, head held high, waving to the roaring crowd. The reporter from the *Berliner Tageblatt* captured vividly the mood at the racetrack.

Only Flying Is Nicer

The history of the feeling of speed is greatly indebted to motor sports. Without car racing the automobile's triumphal march is hard to imagine. Ever since the long-distance run from Paris to Bordeaux in 1895, which Levassor won with an average speed of 24.4 kilometers per hour, racing has taken care of the publicity. Tens of thousands of spectators lined the streets (in Paris in 1903 it is said that one hundred thousand were on hand). Newspapers and, later, the radio spread racing fever all through the country. The sports pages were filled with reports of all kinds of big prizes, whether at Monza, Indianapolis, or even at the Nürburgring race; and every child knew the names of Lautenschlager, Caracciola, and Rosemeyer.

To be sure, engineers were putting the finishing touches on their test machines, and advertising experts were adding luster to their firms' image. But it was racing that popularized the new role model: the athletic, triumphant driver. Who knew, at the time of Lautenschlager, what it took to be a real automobile driver? The congenial coachman and the fiery rider were familiar images, but what the model automobile driver looked like, how he should deal with his vehicle, what his secret behind the wheel was—of that, there was not yet a clue.

Car racing offered relief. Whereas in daily life only fragile vehicles were to be seen putt-putting their way past the chickens in the road, the races presented a pure image of the automobile driver, an image in which all that counted was carrying the day in the competition of speed. Because the excited masses identified so strongly with the idols of car racing, nothing less than a new perception of reality was installed: the drawing of pleasure and superiority in the role of driver, by teasing the limits of both the automobile and one's own fate, so that the world, the tired, old world, flew by and an admiring gaze looked after. Thus did the thrill of high-speed driving become established in the public's fantasies through the spectacle of car racing.

Over the years and decades the fantasies became reality. In a 1971 survey of German drivers in the newsweekly *Spiegel,* 36 percent spoke of the fun they had pushing their cars to the limit. By age, the eighteen-to twenty-nine-year-olds, by make, the BMW drivers especially liked

On your mark . . . Caricature by Bruno Paul in Simplicissimus, *1906.*

to step on the gas.[2] An investigation of German highway drivers by the Del-Berg Institute in Cologne likewise revealed that driving fast was a source of intense experience.[3] Fully two-thirds of those surveyed mentioned a "pleasurable" feeling from driving, that it relieved an "itch" and was somehow "seductive" and "intoxicating."

Driving fast, it turned out, is pleasurable because it entails a risk; it entices with the sweet poison of danger. Driving is bound up not only with the feeling of rising and flying "beyond limitations," but also with the anxious tension of having risked too much. It is precisely in this

2. "Der Deutsche und sein Auto" (The German and his car), *Der Spiegel,* 1971, no. 53.
3. H. J. Berger, G. Bliersbach, and R. G. Dellen, *Macht und Ohnmacht auf der Autobahn. Dimensionen des Erlebens beim Autofahren* (Power and impotence on the Autobahn: Dimensions of experience in driving) (Frankfurt, 1973).

Coupe Gordon Bennett, by Henri-Charles Willems, 1905. From Hervé Poulain, L'art et l'automobile (Zug: Les Clefs du Temps, 1973).

mix of anxiety and pleasure in overcoming it that the "thrill" is rooted, that tickling of the nerves that many experience at high speeds. Just as exciting as rock climbing or hang gliding, driving fast invites one to balance on the edge between power and impotence, and then to enjoy the gratification of not having crashed. Whoever feels the need to overcome hidden doubts and to find self-confirmation in strength has ample opportunity as a little Caracciola. Through the experience of the "thrill," fantasies of greatness long since worn out or given up may be redeemed in real life: adults can again dream, as they did when they were children, of being larger than life. Speed enthusiasts feel hemmed in when they cannot run their cars "full out." For them "real driving" begins only on an open stretch of road. The experience of power sets in because they are able to rise at once above nature ("weightlessness"), the vehicle ("full bore"), and other people ("left behind"). Since one can fully surrender oneself to this power tease only so long as no distrustful looks from a passenger call to mind repressed feelings of guilt, the speed enthusiasts, according to the survey, most prefer to drive . . . alone.

Racing taught drivers to expect from their own automobiles a thrill that, through role identification, they then carried over into their daily practice of driving. Adventure and danger, the spice life acquired with the automobile—this theme underlies, for example, the humorous novel by H. G. Bentz about the world of driving:

> They began to lead a double life. In the first, until now the only, existence they were just a number in some office. . . . They had the feeling of running around a big cage their whole life long. . . . Suddenly adventure surrounded them, opening new chasms, storming unknown peaks, unfolding unsuspected paradises. Suddenly danger was there too, around every curve, before every incline, on the edge of every roadside ditch; it radiated from the trees lining the street, which lured the driver magnetically, as the sirens' songs lured Odysseus. Danger came from the child suddenly leaping onto the embankment, from the tire that blew out, tearing the wheel from the driver's hand like a giant fist; danger from the slippery pavement, from badly marked or wrongly banked curves. . . . Neither pension rights nor major medical insurance were of use when they crashed into another car or wrapped around a tree; only alertness and self-discipline helped, and sometimes not even those.[4]

The Power of Purchased Force

The automobile in particular promised the feeling of omnipotence; of the tools that industrial progress has brought within the reach of individuals, no other represented so disproportionate a gathering of impersonal energy to strengthen weak human powers. When an imperceptible curling of the toes, the scant flex of muscle, suffices to release forces that multiply those of the driver many times over and allow him to take off like a shot, then it is already given that feelings of power will be sought in the automobile. Because the sudden roar of acceleration does in fact respond to the pressure of the toes and the steering wheel to that of the hands, rather than—as, for example, in an elevator— deriving strictly from the machine's automatism, the driver can perceive himself as master of this force and experience it as an enlargement of the self.

In this gaping incongruence between steering and steered forces lies the key reason why the automobile is especially suited as a means of self-reinforcement, as an ego prosthesis for those needy of power.

4. H. G. Bentz, *Alle meine Autos. Herrenfahrers Lust und Leid* (All my cars: A gentleman driver's pleasure and pain) (Munich, 1980), 11.

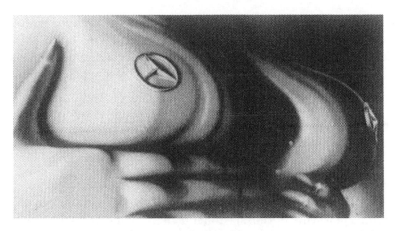

"Frauto" (Car Woman), by Hugo Schuhmacher, 1970. From Hervé Poulain,
L'art et l'automobile *(Zug: Les Clefs du Temps, 1973).*

With the force of acceleration the motor propels the forces of the ego
as well; that is why masculine fantasies and feelings of omnipotence
crystallize about the automobile. A "lame" vehicle that does not jump
off the line when the light changes awakens impotence, even castration
anxiety; where else, expressed in tenths of a second, is the mythical
magic of acceleration values to be found? Now that it is no longer
fashionable to slap up admiration for oneself with one's fists, the car
offers the best medium for powerful self-presentation, from which
even the weakling emerges strong. Indeed, the novelty in this model of
experience lies in the very fact that the feeling of power is divorced
from personal characteristics (aside from purchasing power). Before
the automobile, the experience of power was bound to one's ability to
control, through physical force, prestige, or wealth, the wills of other
people (or animals), in order to draw superior force from them. But
the automobile offers the feeling of power without the consent of
others, and this power no longer has to be earned—or rather, it can
be earned only in the financial sense. Thus experiences of power, even
if completely devoid of influence, become accessible to everyone; that
is why often the average man fears speed limits as if they might take
something from him.

The qualities of the car engine are reflected in the driver's character,
and the choice of engine lends expression to the driver's self-image:

this connection of man and machine is advertising's favored representation. In a commercial for the VW Jetta, the car—the driver all concentration behind the wheel—approaches the viewer with such power that it practically leaps out of the picture. The text explains:

> The new Jetta. Dynamism from the factory. The first look proves its strength of character. With a wedge-shaped body. With wide-angle headlights and integrated blinkers. A low front end and an aerodynamic spoiler. And don't forget: its trunk space—630 liters. Making the Jetta a luxury car that has it all.

Styling, name (Jet-ta!), and the advertising message are supposed to make the car into a symbol of dynamism and strength, indeed in the sacramental sense of causing a literal effect: whoever takes the wheel gains dynamism and strength as a person.

And the driver is truly in need of dynamism and strength, for only then can he win the manifold competitions of the street. Unfortunately, the streets, and especially the highways, have not been swept clean of competitors; on the contrary, in every direction other cars are disturbingly evident, and at every second the striving for power can end in defeat. Because by the nature of the thing all drivers want to proceed quickly for reasons of their own, traffic is oriented from the beginning toward conflict; the intentions of any one individual stand opposed to those of all the others, threatening constantly to collide. Drivers therefore see themselves, whether they want to or not, as pressed into a rivalry, a competition, in which quick acceleration and an ample top speed come in handy. The point is not to be frustrated in one's own desire for speed or be hindered or taken advantage of by the intruders.

Driving, in short, necessitates an egocentric perspective in which other drivers appear first of all as competitors. These competitors call on one both to strive for victory and to fear defeat; driving in the end is experienced as a series of small rebuffs and triumphs. Changing lanes, passing, and successful accelerations are summarized as personal victories, while whoever is cut off, blocked out, or left behind suffers defeat. The desire for victory and the fear of offense call up a mix of arrogance and nervousness that besets most people as soon as they take the wheel. Intensified by the fact that cars are "communication prisons"—their mimetic rigidity prevents them from explaining, excusing themselves, or at least smiling in embarrassment when they squeeze someone off the road—every hindrance becomes a personal attack and begs for retribution.

Aggressiveness in traffic is nourished from an offended love for speed. The following common situation on the highway is all but archetypical: you are in the process of passing when another car appears from behind in the rearview mirror, flashing its lights and scaring you back into the right-hand lane; you would love just to step on the brakes; but you get out of the way, all the while burning to show next what you—pardon, your car—can do. A car with lots of horsepower is the best defense, as well as the best offense; thus the ambitious prefer high-performance vehicles to score a victory, and the successful to avoid having to accept defeat—that is the difference between BMW and Mercedes owners. In any case, as studies have shown, drivers of larger cars feel obliged to navigate at higher speeds than drivers of smaller cars and to change lanes more often and tailgate more closely to clear the way. And they do not take speed limits so seriously, for as the horsepower under the hood mounts, so does the feeling of self-worth: in a survey, 92 percent of Mercedes drivers considered themselves good or very good drivers, while among Volkswagen drivers the figure was only 59 percent.[5] It is the drivers of large cars who have an especially marked need to win and therefore press the speed.

As motorization proceeded, the joy of small victories came at the cost of increasing frustration; the more traffic there was, the more vulnerable became the love for speed. The ideal of the athletic, triumphant driver was actually formed in artificial circumstances, namely, on isolated race tracks; as an ideal, then, was it not contrary to the daily realities of the street? The hidden preconditions of this model of fast driving are empty streets and competitors striving after the same goal. Yet the heavy traffic of today is inconceivably distant from such conditions. Whoever presses for speed here gets only psychological tension, irritation, an elevated pulse—in short, a full measure of stress. Where is the forceful victory when one is constantly threatened with being trapped between the man dawdling in front and the one pressuring from behind? The victories become more costly in heavy traffic, for they occur less often, require greater concentration, and end more

5. B. Biehl, Dieter von Klebelsberg, and Uwe Seydel, "Einstellungen und Verhalten gegenüber Geschwindigkeitsbeschränkungen auf Autobahnen" (Attitudes and conduct relative to speed limits on the highway) *Zeitschrift für Verkehrssicherheit* (Magazine of traffic safety) 16 (1970): 285f.

quickly in a traffic jam; the small defeats become less avoidable, for competitors appear from everywhere and maneuver one into a situation where there are only losers and no winners at all.

The decline of the love of speed cannot be far behind, for every advantage comes at the cost of stress, and acceleration serves primarily to escape narcissistic offenses. Indeed, many feel obligated to use like tactics on the faster drivers; thus more people are drawn into the undertow of speed and the atmosphere of compulsive nervousness spreads. The ideal image of the athletic, victorious driver is transformed into an anxiety-ridden image of the hunted and stressed driver. It is no surprise that speed limits are increasingly experienced as relief, because they defend against offense and outbreaks of sweat. Probably for these reasons, 54 percent of private automobile owners voted for speed limits in 1977, while only 30 percent spoke out against them.

The Sense of Speed
Born of the Belief in Progress

The model of the athletic, victorious driver was a cultural creation of car racing as a sport. It is thus no accident that races came into vogue immediately in the early years of the automobile, that racing had substantial influence on automobile design into the 1930s, and that, above all, the sport left its mark on the cultural definition of driving. Car races cast a spell on the people and filled their spirits with suspense and excitement even as the vital feel of the epoch became more concentrated in the dynamic of mobility. Perhaps no other mass ritual is so perfectly expressive of the turn-of-the-century *Zeitgeist*.

The rise of the industrialized nations, their imperial thrust beyond established boundaries, the murderous competition among entrepreneurs and nations with victory going to the strongest and fittest—the driving forces of industrialism had truly conquered the people. "The rapidity of any given event," wrote Werner Sombart in 1920,

> interests the people of today almost as much as its enormity. To drive in an automobile at "a speed of one hundred kilometers": our time truly dangles that before the eyes as a supreme ideal. And those who cannot themselves surge forward take pleasure in the numbers they read of some speed achieved elsewhere: that the express train between Berlin and Hamburg has cut ten minutes from its scheduled time; that the newest steam leviathan arrived three hours earlier in New York. . . . All the mania for greatness and speed of our time finds its expression in this concept of the record. I do not think it unlikely that

a historian who undertakes in a couple of hundred years to describe the present in which we live will entitle this period of his study "The Age of the Record."[6]

In fact, it was only toward the end of the nineteenth century that the word *record* appeared in the German language, an indication of a deep transformation in how accomplishment and time were conceived. Still reflecting its early meaning, namely a "written note," the word began to be used in 1886 as an official confirmation of the highest achievement in a sport, and only after the turn of the century did it come into use as the achievement or first-place ranking itself. Already in 1913, the economist Joseph Schumpeter accepted the concept as being in popular use—a truly rapid development, and one proclaiming the arrival of a new view of the world.

In sports, where the concept first took hold, a new ideal of movement and achievement became visible over the course of the nineteenth century, which then was aptly applied to motor sports and laid the basis for their attractiveness. Car racing represented the end point of a development that changed the very nature of many sports, indeed was responsible for the creation of modern high-performance sports. Traditional forms of physical exercise had been increasingly modified and then replaced by varieties of sport in which the entire point was to achieve the highest marks in a struggle against centimeters, grams, and seconds, a notion of athletic activity that was in no way common before the middle of the nineteenth century and in many cultures was unknown altogether.[7] Thus, for example, the art of riding, which in early modern times was a matter of perfectly executing required movements at a predetermined tempo, was replaced by horse racing, where the laurels went to the rider who covered a given stretch more quickly than his competitors. Likewise the traditional tests of strength, such as stone lifting, stone tossing, or boxing, gave way to high-performance competitions, in which weights and distances were measured systematically and the victor made his mark by means of a quantitative accomplishment, no longer to be celebrated at local festivals for astonishing feats of strength. This was also the period when the modern timed

6. Quoted in Rudolf Wendorff, *Zeit und Kultur. Geschichte des Zeitbewußtseins in Europa* (Time and culture: History of the consciousness of time in Europe) (Opladen, 1980), 552.

7. This hypothesis is the basis of Henning Eichberg's study of the social history of sports in the last century: *Leistung, Spannung, Geschwindigkeit: Sport und Tanz im gesellschaftlichen Wandel des 18./19. Jahrhunderts* (Stuttgart: Klett, 1978).

sports were born—track and swimming contests—in which speeds are measured in fractions of a second and the victor is expected to better a generally recognized record.

Whatever the shifts in individual areas, the ideal of linear increases in performance was commonly assumed. Whereas in the traditional disciplines of movement spatial-geometric harmonies were paramount and exercises were to be judged according to "gracefulness of form" or "proper proportion," now the orientation toward measured records, toward top performances viewed quantitatively, pressed to the fore. With the first Olympic Games, that world ritual of records, in 1896, the transformation of sports was complete. *Citius, altius, fortius* would henceforth be the rhythm according to which the *Zeitgeist* itself choreographed athletic competitions.

The transformation drew its meaning from the faith that the future will always be superior to the past and that history strides forward in a straight line to ever greater accomplishments. A social melody resounded on the raceways and running tracks, reaching a crescendo at the turn of the century: upward and onward along the rising path of unending progress! The dynamic of movement in the new varieties of sport copied the dynamic of movement in history with which the people of the time were so enchanted. Just as progress leaves the past behind, linearly and irretrievably, through ever more brilliant achievements, so would the competitions over grams, meters, and seconds continually topple the old records. "There is," to quote Lewis Mumford, "only one efficient speed: *faster;* only one attractive destination: *farther away;* only one desirable size: *bigger;* only one rational quantitative goal: *more.*"[8] Upward and onward appeared to be the movement everywhere: electricity made the night into day; the railway defied the mountains and valleys; and the telegraph encompassed the world. The decades from 1870 to 1910 were the classical years of the certainty of progress.

When the future holds all the promise, nothing remains for the present but to hurry. With an easygoing attitude one misses out on the future; acceleration becomes key. This transformation in the conception of time appears clearly in language, the seismograph of deep shifts in mentality; for language supplies the categories with which people take in the world. It is instructive, for example, to see how the mean-

8. Lewis Mumford, *The Myth of the Machine: The Pentagon of Power* (New York: Harcourt Brace Jovanovich, 1970), 173.

From People's Police, *London, 1980.*

ing of the word *tempo* had reversed itself by the turn of the century. Had one spoken to the young Bismarck of the tempo of time, he would not have understood; he would have thought of the military or, at best, of music. For *tempo* meant "in the appropriate measure of time," whether in terms of execution of a command or beat in a composition. Only toward the turn of the century did *tempo* take on the colloquial meaning of "a high rate of speed." Thus language reflects that an "appropriate measure of time" is self-evidently "a high rate of speed"! As in sports, the ideal of any given motion no longer lies in the proper proportions, but in the fastest possible speed.

The notion of speed proceeding straight as an arrow could only arise from the experience of a machine-driven means of transportation, namely the railway. People or horses are always threatened by exhaustion and weakness because of their organic nature; they can be fast only in proportion (!) to their powers. The locomotive does away with this limitation. It bursts the bounds of organic nature, appears to race tirelessly over mountains and through valleys, hindered by neither its own corporeality nor topography. One of the first reports concerning rail travel makes this distinction plain. "The fiery dragon in front," wrote Friedrich von Raumer from England in 1835,

> snorting, groaning, and roaring, until the twenty cars are fixed to its tail and it moves them, light as a child, over the level tracks at an

extreme rate of speed. A path has been broken through mountains, valleys have been raised; the dragon throws sparks and flames into the night of the arched tunnel. But despite all the violence and despite all the roars, a human being turns the monster to his will with the touch of a finger.[9]

Like a projectile, as contemporaries express it, the train shoots through space; the passenger sits calmly as forests and villages rush by outside. The distance from departure to arrival fades to a mere intermediate space, which makes itself evident only as a bother and the point of which is to traverse it as quickly as possible. With rail travel, the notion of speed straight as an arrow gained clear form in the mind and could become a (machine-driven) ideal that cast all human motion as well as history in a new light.

Speed, this annihilation of intermediate spaces, brought distant goals within reach. "The North Sea breaks on my doorstep," was Heine's famous commentary on the railway from Paris. The railroad became an archetype of progress, because it seemed to correspond so completely to the basic desire of progress: that there be no limits on the way to a richer and better world. Yet space and time, exertion and duration, always stood in the way of this longing, were, indeed, practically the basic forms of hindrance, which is experienced with more aggravation the stronger the longing. The railway promised to open the borders of space and time; thus the breakthrough into the future, into the paradise of industrial wealth, seemed a decisive step nearer. Simultaneity, omnilocality, and the annihilation of distance would, in the utopian view of things, be the consequence of the belief in progress. Indeed, this hope was the sign of the time, and everything appeared to be pointing toward a breakthrough into this utopia: with coal, oil, and electricity, were not the miraculous tools at hand to allow, with the railway, automobile, and telephone, geography to be vanquished?

At the car races, whether peering from roadside at the automobiles speeding by or hearing the crowd's jubilation on the radio, the public participated in the cult of straight-as-an-arrow speed. The excitement and desires of the time were celebrated in the sport of racing. The athletic, victorious driver was not only a hero of the streets, but a hero of continual progress as well. It only makes sense that futurist artists would raise the race car as the symbol of the epoch's essence: "We de-

9. Quoted in Riedel, "Vom Biedermeier zum Maschinenzeitalter," 111.

clare that the mastery of the world has been enriched by yet another beauty: the beauty of speed. A race car, its body adorned by great pipes the equivalent of snakes with explosive breath . . . a roaring automobile, which appears to roll on artillery shells, is more beautiful than the Nike of Samothrace." [10]

The Bias Toward Speed Machines

The automobile has various advantages: it allows one to travel comfortably, to be sheltered from wind and weather, to transport cargo, or . . . to triumph with speed. To which of these expectations the automobile is particularly suited, however, is not given in advance, but determined by the priorities of the manufacturer and buyer. The expectations that are most prominent in the eyes of the customers and designers—in terms of engine technology, design, and equipment— have had a decisive influence on the development of the modern fleet of automobiles.

The history of the automobile is full of deviations, of paths not taken and then forgotten. Designs were made, experiments conducted; but by and large, a special class of expectations determined the main path that development would proceed: to ever higher speeds and performance capabilities. No less a figure than Carl Benz attempted to deflect this course of development as early as 1901 when he refused (in contrast to the Daimler Mercedes) to trim his automobile to the requirements of speed. "Among the dangers to the development of automotive concerns" he counted in particular "the new craze for outdoing oneself at the races with ever-increasing speeds, to compete with express trains and thereby frivolously risk the lives of drivers as well as of the people moving about the streets." [11] Benz, however, proved unable to win his way and was driven from his company when a French team of engineers was forced on him. He and a few later opponents of the trend simply could not fight off the draw of expectations, which has permanently stamped the development of automotive technology: high top speeds and quick acceleration as the first priority in automotive design.

Of the multitude of possible developmental paths, the royal road to the speed machine was taken: large, high-performance vehicles and,

10. Quoted in Wendorff, *Zeit und Kultur,* 556.
11. Paul Siebertz, *Karl Benz. Ein Pionier der Verkehrsmotorisierung* (Karl Benz: A pioneer of motorization) (Munich, Berlin, 1943), 181.

later, small, spirited cars became the models of automotive design. Thus technological development followed a cultural project in that it arrived at speed and acceleration. Car racing not only paraded the ideal of the athletic, victorious driver before the masses, but it also decisively influenced the technical conception of the automobile through its first decades. Engines and associated systems, chassis parts, and bodies, for example, had to withstand stress tests on race cars before they were introduced into daily use. Moreover, after the 1920s automobile design was quickly beset by a technological rigidity that allowed the cultivation of speed to become the dominant developmental line, and cars since then have remained unchanged in their basic structure. Aside from a few innovations—which were actually introduced from outside automotive technology, from metallurgy and petrochemicals, pneumatics and electronics—technical inventiveness was exhausted in suiting engines, chassis, and bodies to speed. Cruising speed was to approximate top speed, more horsepower was to be made available from engine volume, and, most important, the vehicle needed to lose weight. Ever more "spirited" automobiles offered at an ever-decreasing price: that is the car manufacturer's creed.

It is no surprise that between 1959 and 1979, the years of the great motorization boom, the proportion of vehicles on German streets with engine capacities exceeding 1,500 cc rose from 13 to 51 percent; the auto-mobilization of society was clearly marked by a trend toward high performance. How thoroughly this model also dominated research was first documented in a federal research program (at a cost 110 million marks) on the automobile in the year 2000, in which the best engineers were called on to square the circle: to conceive an environmentally sound, economical automobile—that still had the power to cruise, unloaded, at a speed of at least 160 kilometers per hour!

Even today, when cars are driven mostly in city traffic, their design characteristics are oriented to high-speed, long-distance travel; with their powerful engines, their streamlined bodies, and their high-speed suspension systems, they are as suited to the delays of city traffic as a chainsaw is for cutting butter. Moreover, is there a single private automobile on the market today whose top speed is even roughly proportional to the speeds allowed or recommended on streets and highways? It is not the supposedly rational spirit of the engineers that is at work here; rather, their engines are monuments to the love of speed.

Secure in Comfort

"My dear," wrote a young French aristocrat to his lady friend in 1899,

> I have been keeping, despite a thousand unpleasant details, one of
> those modern gasoline vehicles for two months now. I'm not dissatis-
> fied, but meanwhile I have spent only part of the time in the passenger
> compartment, a larger portion of it underneath. You will find it impos-
> sible, dearest, but I am surrounded by the most penetrating odor of
> gasoline you can imagine. And even my hands—they resemble black
> paws. . . . As I returned home from the repair shop yesterday, I invited
> the Countess Mantua—you will recall the elderly lady—on a little
> drive; when I disengaged the motor, the shudder of the idle actually
> caused her denture to fall in her lap. She has been avoiding me since.
> And please forgive me the poor handwriting; as of yesterday I have
> carried my right arm in a sling, which was bruised by the crank as I
> started the machine—a misfortune, as I hear, that one must endeavor
> to accept as part of the whole.

The young aristocrat slid from catastrophe to catastrophe. Like
Laocoön of old with the snakes, he struggled with the object's pranks;
the ecstasy of the motor was purchased at the cost of oily fingers,
frozen body, and bruised limb. But how the lot of the drivers has
changed in the meantime! Could our horseless carriage pioneers have
imagined how the automobile has come along? "The fascinating aspect of the new Audi 200 Turbo," pronounces a consumer report,

> is its overall superiority. The engine does its work in a supremely culti-
> vated fashion. Even at speeds of 220 kilometers per hour, chatting in a
> discreet conversational tone remains possible. The motor never be-

Breakdown adventures in the early days. From Jacques H. Lartique, Photo-Tagebuch unseres Jahrhunderts (Lucerne: C. J. Bucher, 1970).

comes really loud, and at the same time noise from the wind remains subdued even at top speed. The chassis is first class, for it allows a dreamily secure drive on the straightaways and almost unparalleled handling on high-speed curves. . . . The list of standard accessories fills several pages, running from an automatic warning system through an electronic tachometer to power seats, power windows in the front, central controls, light metal wheels, and adjustable-beam headlights, as well as the newest antilock braking system.[1]

Which cultural track has automotive technology picked up here?

In the Care of Instruments

The taste for comfort, it can be said, stems from the living room of the nineteenth-century bourgeois family. When Prince Pückler wrote his widely publicized letters from London in the 1820s, he made it clear that he was impressed by the new luxury he found in the houses of the English middle class. He extolled the English bourgeoisie for not having dissipated their surplus, like the aristocracy, in ostentatious splendor, but rather investing it in objects of utilitarian convenience. He summarized this lifestyle, describing the rugs and drapes, mirrors, chairs, and

1. *Frankfurter Rundschau,* July 2, 1983.

toilet facilities, in a single word that would thenceforth be quite popular: comfort. It was with this concept that the prosperous bourgeoisie could distinguish itself from the nobility, which was accused of wasting its surplus wealth unproductively in elaborate pomp, and justify, in contrast to "useless luxury," a new, "rational" kind of spending: the creation in the private sphere of an agreeable and protective atmosphere. Such an undertaking was in bourgeois opinion much more useful than dissolute luxury. The goal of bourgeois industriousness, after all, consisted in the realization, through the production of goods, of happiness here in earthly life. Boisterous parties and high-value luxury articles for individual use were suspect in the eyes of the accumulation-minded bourgeois; for them the only legitimate consumption was one that created a comfortable domesticity.

In any case, prebourgeois residential arrangements were not conducive to the pursuit of a private life. Cultivation demanded the creation of distance, the dampening of both the hubbub of the people and the exuberance of emotion; it also called for shielding the discharge of bodily functions from the eyes of others. Private spaces, bedrooms, and toilets were required for the creation of an intimate sphere, in which individual inclination and cultivation of the personality could flourish. This transformation—legible in changes in the floor plans and interior design of dwellings—was driven by the same intention, that is, the creation of a protective zone of distance around the individual. Model behavior now came to be defined by a striving after "higher" things, and with it a sensitivity to the intrusions of daily life emerged; only when noise and sweat and strife could be held at some remove could one advance to cultivation of the inner life. All kinds of arrangements were introduced to cushion the ego from all sides: one no longer received visitors in bed, as the nobility of the eighteenth century had done, but in the living room; domestics (as opposed to servants) saw to daily welfare; rugs and curtains muffled noise; new stoves equilibrated the variation in temperature; upholstery protected against hard spots and corners; and indoor plumbing spared one those inconvenient strolls out-of-doors. Well-being consisted of protection. The new arrangements were supposed to provide for islands of undisturbed peace, where no sudden bodily sensations, no outbursts of emotion, no noisy strains interfered with the striving after higher goals. Comfortable domesticity provided protection against the tedious annoyances of daily life.

While at the beginning the domestics were there to shield master

A DREAM OF PROGRESS
BY ERNST SEIFFERT, BERLIN

(from a diary)

October 19, 1963

It is still most peaceful on the level ground, for on the causeways there is only the crowd of motorcars, while the former footpaths have been reserved for the pedestrians and, thanks to the moving sidewalks, make a more harmonious impression than in earlier times. These moving walks are an ingenious arrangement. They lie next to each other in three parts, thereby providing three different speeds: ten, twenty, and thirty kilometers per hour. This system of moving sidewalks runs through the whole of the central city. The intersections are outfitted with rotating disks, so that with a simple step one can enter the desired side street. The powerful lines of the elevated railway extend powerfully above the streets. The familiar open elevators run at an incredible speed up to the stations; for a more comfortable exit from the elevated, slides slanting down from the top have recently been constructed and have proved first rate. Above the elevated railway moves the airship and aviation operations. Here there has been such progress that for some time now one might have had the idea that national borders have been declared null and void, because it proved impossible to maintain the formerly common customs system. Here in Berlin every building in the inner city has its own rooftop garden, with the required landing pads for airships and airplanes. Large central stations have also been provided for the air busses; one ascends to them on elevator systems, as to the elevated railway. Likewise, elevators and slides lead down to the subway, the comprehensive network of which was completed a year ago for all of the traffic arteries of Berlin and which is now able to handle the significant increase in use, since it has four tracks nearly everywhere, often—on the main routes—even with two or three tracks layered one over the other.

Be that as it may: what has been accomplished in the last few decades is beyond conception. Who would have believed in 1940 that the plague of traffic jams would finally, finally be completely overcome? Had a person proposed polished asphalt or certainly the newest hard-rubber streets back then, he would have been laughed at. How wonderfully glass-smooth the streets of today are! How fine the effect of the napped hard-rubber surface over the moving sidewalks, how fortunate that the oil paints and lacquers have been replaced by chemical coloring treatments of the steel itself, which cannot be damaged, never fades, is resistant to water, even acid, and does not rust or even get dusty! I read in a book published in 1912 that machines had robbed the world of character and made it too noisy. Such a judgment could only be made by one who had not yet struggled through to the stage of great proficiency; had

Vision of an elevated pneumatic train system to solve future transit problems. From Jacques H. Lartique, Photo-Tagebuch unseres Jahrhunderts *(Lucerne: C. J. Bucher, 1970).*

one foreseen today's condition, the judgment would presumably have been otherwise. Is there a more gigantic, more vital image than that of the big, modern city of today? Where do we find the same bold lines of streets and transit trains, where similar rhythms, where the same joy in the strong pulse of life as here?

Das Neue Universum (The new universe), August 1913.

and mistress from the toil of everyday life, after the turn of the century, as domestics became scarce, expensive, and refractory to boot, an increasing number of technical appliances took over the task. Vacuum cleaner, gas stove, central heating, electric mixer, or elevator—all converged on the point of making daily life run more smoothly and, above all, of eliminating physical sensations and exertions. Thanks to "appliancization," comfort was soon brought within the reach of everyone; the desire to surround oneself with appliances that did away with dirt, sweat, and toil became a general good.

Everywhere the burgeoning appliance technology was changing the gestures of daily life, all in strict accord with the aim of getting as much as possible accomplished with the least possible involvement from the user, setting an army of mechanical slaves into motion at the touch of a button. Recalculated in terms of human performance, the average American in 1978 commanded the labor of about one thousand energy slaves every day. The hidden agenda of comfort technology consists in relieving, indeed shutting off, the body and head, allowing the everyday world to surround people like a tailored coat: a perfect fit and without a trace of its presence.

Turn-of-the-century sketches depicting the technological-utopian future outdid each other portraying the humans of tomorrow as remarkably passive beings. They glide toward their destinations, for example, on rolling sidewalks with roofs, seated, if desired, with no need to move a muscle. In the dining room, they are served a multicourse meal by means of lifting, pulling, and lowering mechanisms of every conceivable type; no one need even rise from the chair. Both sketches offer a glimpse, one could say, of the metaphysic of comfort: convenience machines provide a technological answer to the venerable question of how to overcome human frailty. "Con-fort," after all, means, in its original sense, "to fortify, to console." Do not the appliances take over that which once found only religious meaning, namely, the transcendence of human weakness and mortality? Perhaps it is along these lines that we may discover why "the loss of comfort" is often experienced today as a fundamental threat: robbed of the protection of appliances, we feel ourselves delivered up to our own frailty. Perhaps, indeed, we are all the more defenseless today in that technological development, rather than saving us, has in fact caused our moral-cultural capacity to deal cheerfully and easily with our limitations—to be able "to suffer" them well, as the old expression paradoxically puts it—to atrophy.

Cozy Regression

The technical development of the automobile was also oriented on the tacit assumption that human beings are defined by their need to be sheltered and relieved of burdens, to be guarded by technology against the influence of wind and weather, against bodily exertion and mental effort. Vehicles become the repository of their drivers' desires for comfort, which appear to stem equally from simultaneous wishes for regression and omnipotence. One sinks in so cozily, almost as if into the sheltering womb, surrounded by humming technology and unmistakable comfort, while at the same time indulging a feeling of power and force rarely experienced in life otherwise. Racing in the night through a lashing rain, one feels at once protected and strengthened, and lulled, too, by cozy warmth and rhythmic music. "Quickly and tirelessly," a weekend traveler enthused in 1929, "the windshield wipers do their work, clearing the view and delivering the dripping world through their arc. It is both mysterious and familiar to sit in the snug warmth of the car, with soupy fog all around, and snow and driving rain." The wish for security forces its way to the surface, paired with feelings of triumph, for it takes only a little pressure on the pedal to flee even layers of adversity. "Meter-deep snowdrifts prohibit forward progress, ever farther the hood digs its way into the porous bank. The fenders disappear; we sit ensconced in the treacherous snow, in ice, in slush. The tires dig fitfully deeper, gain no grip. I pause briefly, shift into reverse, attack again: the car thrusts forward, screaming RPMs . . . and wins."[2] The theme of gratification here—"the protective shell against wrongs from without"—is by no means bound exclusively to confrontation with the forces of nature: in the car one can seal oneself off from all kinds of stress. Enough of the boss's nagging—one sinks down into the cushions, pulls the door closed, inserts a favorite cassette, starts the engine: the car glides obediently out of its narrow space. In the daily commute to and from work this experience plays a particularly significant role. *My car is my castle.*

It is no wonder that the automobile is especially suited to a charge of narcissistic feeling, for it, like scarcely any other apparatus, represents a man-machine system in which, from the toes through the hips to the eyes, all senses are linked to the machine. Drivers must extend their region of sensory attention beyond themselves to the whole ve-

2. *Der Motor-Tourist,* 1929, no. 1, 8.

Turn-of-the-century "comfort." From Jacques H. Lartique, Photo-Tagebuch unseres Jahrhunderts *(Lucerne: C. J. Bucher, 1970).*

> Always ready! The more the automobile becomes useful for every occasion, the clearer it is that special driving clothes are absurd and inappropriate. Today one sits behind the wheel, presses on the starter, and drives off. And since one no longer needs to perform manual operations, which could lead to dirtying the clothes, it is also insignificant what kind of clothes one has on. The days of getting all bundled up to take the wheel of an automobile are definitely past.
>
> *Elegante Welt* (Elegant world), 1926, no. 15, 21.

hicle and synchronize its mechanical functions with bodily movements, indeed, even internalize the ensemble of "man and machine." Because of this symbiosis, it is very easy for human sensuousness to become intertwined with the functions of the technology, thus calling up that peculiar emotional mixture of regression and omnipotence. It is cozy, this tension between security and the thrill of power, between uterus and phallus.

Deriving from the incongruity between exertion and effect so typical of the automobile, the tension unifies the dynamic (the deployment of a powerful force) with the passive (a flexing of the toes). Given such an interweaving of feelings, the advertising industry need not stretch far to symbolize the combination of powerful beauty and ease of operation in a sexist manner. On a billboard amid the glitzy Parisian nightlife at Pigalle, a Rover 2600 ("also attractive undressed") is depicted as the beautiful and explosive, but still obedient, female. The car "excites with its elegant shapes," "repays a look under the hood," "opens its interior invitingly," and "wins many lovers." And, because "it handles so easily," you can "let yourself be spoiled" and "speed off to someplace restful" "to enjoy it all alone." No wonder 35 percent of male drivers—though only 16 percent of female drivers—never let anyone but themselves take the wheel of their cars.[3]

Replacing Body and Brain

It has been a long time since the first horseless carriages were taunted in Munich with the jingle "An automobile is a wagon that won't." At least "since the triumph of the American car," as Eugen Diesel expressed it in 1933, "feminine demands have been incorporated too extensively into automobile design." Earlier,

3. "Der Deutsche und sein Auto," *Der Spiegel,* 1971, no. 53.

the machines ticked and knocked a little differently than today; they sucked noisily through the carburetor, and shifting gears in the box was a different art than the one we know, with the synchronized ball-and-socket transmissions of today. After a few thousand kilometers the engines deposited carbon on the cylinders and did not have the all-embracing and yet gentle reserves of power, the rapid get-away power of today's machines. Cranking the engine and mounting the tires were indispensable arts; the rubber felt different and smelled different. The odor of lightweight gasoline suffused the car, oil probably dripped from the bottom, and on high passes the radiator boiled over.[4]

Automobiles continued to change their appearance in the following decades: the engineers saw to it that drivers had increasingly little to be aware of, increasingly little to do, and increasingly little to be responsible for. They have devised countless ways to relieve drivers of burdens, and with an array of features, to render any spontaneous activity on their part largely superfluous. Control over driving was increasingly shifted from the driver to technology. To compensate for the loss in responsibility, a "cockpit," adorned with lighted gauges and control panels, was provided, together with a living room–like interior that lacked neither built-in stereo units nor color-coordinated furnishings. The tamer of the gasoline monster has become the pampered inhabitant of a comfort machine.

Not even the smallest unpleasantnesses need impose themselves on the body any longer: finely adjustable heaters make cold feet a thing of the past; adjustable seat inserts pamper the lower back. Even hand movements are becoming superfluous: auto-reverse takes care of the need to turn over music cassettes, and mini-wipers wash the headlights. Ultimately even one's own attentiveness becomes increasingly unnecessary: a warning system reminds the driver of lights left on, and even quick braking becomes child's play with antilock systems. Technology offers itself up like a piece of clothing, ready-made frictionlessness becomes the name of the game.

Once sensation and gestures have been stilled, the comfort technology boasts that it will do away with thinking. Away with burdensome questions: Am I already on the reserve tank? How is the water level in the battery? Or, Is the trunk locked? For an ingenious control system is always there thinking, making itself discreetly known with blinking lights or buzzers. "The feeling of security," an automobile critic enthused,

4. Eugen Diesel, *Wir und das Auto. Denkmal einer Maschine* (The automobile and us: Memorial to a machine) (Leipzig, 1933), viii.

is even transmitted acoustically. If the key is still in the ignition and the door is opened; if the engine is running and the seat belt is still hanging beside the seat; if the key is removed and the headlights are still on—there follows immediately an urgent buzzing or clicking that gets the passenger back in line.[5]

The protective drive of comfort technology has finally penetrated the use of will and thought; appliance technology watches over us everywhere, the caring mother from the beautiful new world.

One hundred and fifty years ago the dawning bourgeois desire for independence led to a striving after comfortable furnishings. Today, after a long history of technological relief for both body and brain, it appears that the desire for independence has turned against comfortable overprotection. Young people especially have become infected with a culture of the unfinished: particle board has achieved a high regard for beds, tables, and shelves in their apartments, while coats and jackets are bought used at flea markets. The use of traditional materials for repairing farmhouses, too, indicates that technological overprotection is losing favor. When the supply of comfort has exceeded a certain limit, the probability rises that what once was welcomed as shelter comes to be experienced as a disabling prison. Not as a prison of oppression, but as one of understimulation—for it is inherent in the essence of the comfort technology that human senses, capacities, and responsibilities lay fallow. And then this quality, the contrary desire for adventure, for effort, for challenge, is released. With less friction in the world at hand, the greater is the attraction of difficulty.

To seek out resistance in order to test and confirm the self is possibly the motivation behind both the adventure vacation and vandalism, both traditional handicrafts and mountain climbing. Too many comfort machines offer relief to the point of understimulation and begin to trivialize human being. From this experience grows a disdainful or careless attitude toward more comfort, even if the desire for the natural gets absorbed by surfboards, white sneakers, or Scandinavian-design furniture. In any case, an unrelenting increase in comfort creates a dilemma for the society of surfeit: because it makes available such an abundance of comfort, it must struggle against an undercurrent of discontent resulting from people's being reduced to button pushers. The twin of satiation is boredom.

5. *Die Zeit,* July 31, 1981.

A Flood of Novelties and the Hunger for Improvement

No language is as international as the language of automotive design: most people can judge the age of a car with a quick glance. Just a silhouette can call up memories of past years or even epochs, for automobile distribution proceeds uniformly, year by year. Contours or color, equipment or engine technology—what appeared yesterday with the polish of the new, was the object of ravings by advertising people and car reviewers, is today ignored by friends and neighbors and starting to show the patina of age. The new models come out in annual cycles, proud representatives of the achievements of progress; then, slowly growing old and ugly, they sink in the next ten years to the bottom of the automobile fleet, landing finally in the crusher at the salvage yard. This change from novelty to commonplace to wreck is in no way merely a material process of aging; it is also a cultural one: new perceptions and needs make their appearance, then pale to the commonplace or, having become jokes, withdraw after a suitable period into the museum of bygone dreams.

To partake in this history of tastes, one must enjoy a certain proximity to the public discourse on the automobile; otherwise one can only, like the character Lenz in the story of the same name by Peter Schneider, quietly wonder at the trifles that are honored with the laurels of the new:

One afternoon Lenz walked through the shopping district of the city. . . . He regarded the offerings in the windows. He was amazed that every month new cars, fur coats, shoes, televisions, evening dresses, and suits were put on display there. . . . The changes that had taken place in the last two or three years appeared to him ridiculously minor, scarcely visible with the naked eye. He stopped in front of a VW showroom. He saw that the essential components of the VW hadn't changed at all: it still had the same form, four wheels, two doors; it had become neither bigger nor smaller. But at the same time something had changed. He didn't understand the significance of the lines that he noted in the contours of the fenders and the windshield. He looked at the passersby, who, like him, were now standing in front of the window and looking at the VW on display. He heard remarks that described the car's novelties. Referring to invisible details, the observers discussed the changes in the new model's engine, which was itself not visible. They compared the new engine's cylinder capacity with the cylinder capacities of older engines; they spoke of changes and reinforcements in the body, in the suspension system and the chassis, of a revolution in interior ventilation, and they maintained that the car runs and runs and runs. Lenz inferred from the description that great changes must have taken place. He determined that the changes that seemed to him comparatively unimportant were perceived by most observers as great and decisive.[1]

The Creation of Scarcity

"Nothing should stay the same"—this saying could well serve as the permanent slogan of the International Automobile Show. Indeed, it would be blindingly honest, for the saying hits exactly on the manufacturers' self-admiration and the chummy commentaries of automotive journalists bending appraisingly over the newborn products. "Nothing should stay the same": every new model that rolls off the assembly lines, all polish and sparkle, for presentation to the astonished public casts a shadow over all the models of yesterday and the day before. Each one pushes the others onto the steep incline of depreciation, as even the new model already bears the mark of the past, at least in the eyes of design engineers, who are busy experimenting with the models for the day after tomorrow.

As with all commodities in an advanced consumer society, automobile offerings undergo uninterrupted replacement. What graces the display rooms of automobile dealers today will be freshened up in two

1. Peter Schneider, *Lenz. Eine Erzählung* (Lenz: A story) (Berlin, 1973), 31f.

to three years with new body features or technical variations; in five to six years the basic model will be presented in entirely new dress. After eight to ten years the once proud embodiments of innovation will at best be found begging for a fourth owner in the back of a used car lot. Although it is technically conceivable to design cars that last twenty to twenty-five years, today they depart their lives in traffic after only ten or eleven years on average. Moreover, their life expectancy declined considerably between 1955 and 1972: a car built in 1955 could look forward to 13.6 years of active use, while one built in 1972 could count on only 10.6 years[2]—even though the number of kilometers driven per car had dropped by 14 percent. Of course, these figures cover the aging process only in the narrow physical sense, where weaknesses, especially in the body, point the way to the junkyard; they ignore the psychological obsolescence of automobiles, that cooling of fascination that begins much earlier, usually as soon as the spotlight hits a new generation of fresh models. The extent to which these nontechnical factors influence the life expectancy of new cars needs no other proof than the peculiar fact that, in Europe, cars last almost twice as long in Denmark (15.4 years) as in Switzerland (8.8 years), with Germany in the low middle (10.7 years).[3]

Naturally, the rhythmic arrival of novelties keeps turnover brisk. The automobile market, after all, has been operating for fifteen years along a saturation curve: what is produced can be sold only if replacement demand remains lively. If approximately 95 percent of new cars are sold to people who already own an automobile, producers must clearly keep the time span of first ownership as short as possible, if they are to move their product at all. Under such conditions the policy of offering novelties aims at spoiling the favor new car buyers find in their present cars—even if they are still running excellently—by tuning their desires to brand-new models.

Where a market approaches its limits through expansion, the rate of turnover becomes important. Therefore, life expectancy and, even more, a car's staying power play a decisive role in keeping the assembly lines from grinding to a halt. Since the beginning of the 1960s, as the

2. Peter Helling, *Das Problem der geplanten Obsoleszenz. Dargestellt am Beispiel der Automobilindustrie* (The problem of planned obsolescence: The example of the automobile industry) (Diss., Freie Universität Berlin, 1979), 59.
3. Jean-Paul Ceron and Jean Baillou, *La société de l'éphémère* (The ephemeral society) (Grenoble: Presses Universitaires, 1979), 40.

stratified mass market in Germany made a broad palette of product offerings necessary, product innovation became an end in itself, no longer a mere means of increasing the use value of the vehicles, but a goal that helped to maintain production capacity, growth, and full employment. For this reason the use value of the innovation can be negligible, as long as it carries a proper economic value. In a society that no longer creates jobs in order to produce, but produces in order to secure jobs, the flood of novelties is water on the mill of the production apparatus, which turns out ever more commodities with ever less use value. Because retooling for a new model devours enormous sums of money, even modest progress can be very costly; as a result, the standard of living becomes increasingly "expensive." Inflationary tendencies are inherent in such superfluity. Soon it all begins to resemble the progress of running in place.

Such an economy needs a public that loves its products fleetingly. Consumers are obliged to go along with the novelty cycles and let their aspirations follow the path shown them by the production apparatus: welcoming new products and rejecting the old. Whoever resists, sticking by venerable virtues like economy and solidity, sins against the moral order as it is incorporated in the operating laws of the advanced market economy. A production apparatus addicted to novelty is anything but ethically neutral; indeed, it can be truly understood only when it is conceived also as a moral institution—but a moral institution whose claims come in the garb of objective lawfulness, in the name of competition and progress. In the roundelay of novelties, a message makes itself heard that acts unremittingly on the self-image of consumers: you have no notion of your real needs; what you have is no longer worth anything; go on in and buy yourself some unruffled happiness.

Teachers and ministers might trouble themselves with virtues, but the ceaseless stream of novelties pronounces more effectively on good and bad: it preaches the virtue of easy separation and disdains the ideal of loyalty. The ways of the attentive are antiquated, pride in possession a thing of the past; instead negligence, satiety, and cheerful change are called for. It is easy to discern the throwaway mentality in the culture of the obsolescence economy. The ideal consumer is one who, even in the face of fire and flames, loses interest right after a purchase and directs his desires toward another object. The policy of novelty must in any case speculate on such an orientation toward life, because it requires that the symbolic consumption of products proceed faster than their actual consumption, that desire be written off more quickly than the investment. Long before a car loses economic value, its symbolic value is supposed to plummet. The destruction of the meaning in which the car is wrapped occurs far in advance of the destruction of its tin body.

Novelty demands this advance because, in utter arrogance, it presupposes superiority. It is not the production date as such that makes a novelty of a commodity, but the assumption that it embodies the current state of technological progress. What comes later is likewise better, in that time and progress march ahead to the same beat. The attraction of the new is a function of the image of endless progress, which marks the precise moment when the future overtakes the present and, more important, the past. Had this image not become a daily tenet of faith, despite all the historical evidence to the contrary, everyone would take such commercial slogans as Daihatsu's for absurd:

> Daihatsu. The new new new car from Japan is here. The new new car from Japan is much more than simply a new car: Daihatsu is more economical. Is more comfortable and has more extras. And costs less than a simply new car. If you are buying yourself a new car just now, then go ahead and buy a new new car: Daihatsu.

As devoid of content—in fat bold print—as this advertisement is, it would be utterly senseless if it could not invoke the mythical consensus that the newest of the new is also the best of the best, the tip of the arrow of progress.

In the atmosphere of significations surrounding consumption, the eighteenth- and nineteenth-century transformation of human sentiments lives on: fear of the new gave way to desire for the new, and what was already known shrank to irrelevance in comparison with the expectation of the future—unknown, but nonetheless superior. This image of history underlay the praise lavished on novelties, which fed off the certainty that history constituted an inexhaustible stream of innovations. This image is reproduced today by a system of production that is addicted to novelties. The whole liturgy of design changes, advertisements, automotive reviews, and car shows celebrates with utter self-confidence this version of history as a path of endless improvements. Only in the context of this belief system can the new gain the requisite trust to supersede the suspension system, trunk space, and corrosion protection that is already available and bring into being a

"The world has a new and better car: Nash." Advertisement from Die Dame, November 1928.

world of commodities that accommodates all desires in an ever more streamlined fashion.

The aim of a new product is to make all existing products of its type appear deficient. A commodity that once satisfied is supposed to appear bad; the new draws its shine from devaluing the existing. Thus praise of "the better" is usually tied up with a redefinition of deficiencies. The ad for the new Fiat Ritmo in 1978, for example, clearly attempted to spoil the image of old cars of its type in the eyes of their owners. On the left is a photograph of the new Ritmo, and on the right just the outline of an unknown car. Underneath the Ritmo ("an unusual automobile"), five features are listed that are supposed to make the new car stand out over any other in its class: five gears, noise protection, a large interior, and so forth. But underneath the nameless outline the reader is confronted—parallel to the five features—five times with the dismissive question: " . . . and your car?" The strategy is obvious: by being presented as a car that provides the five characteristics of the new state of progress, the Ritmo renders the old car (and its owner) deficient. "Your former favorite," the ad says at the end of the inquisition, "is it still, after the comparison with the Ritmo?"

The shadow of deficiency is supposed to fall over the familiar old car, making the new one appear more starkly in the light of perfection. Desires for possession are transferred onto the new car; they cannot be satisfied in the usual way now, and in fact the chances that they will be satisfied at all have diminished because they are now identified with the innovations of the new Ritmo. Comfort, security, quiet, and a beautiful shape are no longer obviously and generally available in many cars of this class, but have become scarce goods that can be had only with the new Ritmo. Novelty, by appearing as better, diminishes the chances for satisfaction, because it devalues the familiar and presents itself as the royal road.

This industrially created scarcity sees to it that the desires of the car owner are repeatedly overtaken by indifference and, through periodic changes, are turned toward other objects. As part of the plan, the advanced industrial system aims to keep the relation between needs and commodities unstable and terminable at any time. "Satisfaction subject to recall"—this proviso really should appear at the bottom of every sales contract, for it refers to a characteristic quality of industrial commodity culture—one that distinguishes it from many other cultures and, indeed, represents the very reason for the dissatisfaction of consumer society. It is no wonder that needs never reach a limit, because the scarcity spiral is constantly attaching them to new objects. The endlessness of needs, that famous fundamental assumption of the modern economy, is the subjective side of the endless stream of industrially produced novelties.

The innovation strategists benefit from the plasticity of people's needs and the way clever commodity packaging can call up countless subordinate and accessory needs. How strange it would be for an automobile pioneer from year one to see the "need" that the Ritmo promises to fulfill: "The Ritmo is quiet inside too. At cruising speed you can hear the stereo cassette deck as well as in a luxury sedan. You hear Mozart instead of the motor." Mozart instead of the motor— how much more can desires be specified and refined? A single commercial message for the Ritmo divides the need for a good car into some sixty subordinate needs, from the body-pampering seat and the especially fresh fresh air to the beautifully styled steering wheel and the plastic mudguards. The Ritmo claims to have optimized all of these needs and to be ahead of all other cars in its class, setting the spiral of scarcity in motion for a multitude of needs. Satisfaction is subject to recall, as is a whole series of subordinate needs as well: a multitude of desires have been dipped into the waters of symbolic destruction and resurgent attraction. What drives consumption is the attractiveness of small improvements, the persuasiveness of small differences: a couple of centimeters more here, a couple of seconds faster there; a more pleasing contour here, a new instrument panel light there. Considering all the materials and capital, all the workers and industries, all the social resources that are mobilized for the production of small differences, is it wrong to say that the policy of novelty underlies the most comprehensive and sophisticated ritual of modern society?

Georges Bataille once said that one can identify a society by the form of its expenditures. While in other cultures the surplus is spent on, say, golden altars, ceremonial celebrations, or pyramids, the surplus in modern society is used for the continual advance of production and consumption. The economy, material production itself, has become the religious agent, the instance that creates meaning. The superior commodity dominates the attention of society and is the point around which social forces crystallize; it has something of the status of an identity-bestowing idol, a fleeting one perhaps, but nonetheless dominant.

Why does the small, innovative difference attract so much attention? Why is the new seen as something worth striving for? The Ritmo commercial is not lacking in answers. The Ritmo, it says, "brings a fresh wind to the mid-sized compact." Whereas "a certain conformity" has become evident in the last few years, the Ritmo is simply "other than the others." The car's new characteristics impart the feeling of rising above conformity, for "all of these advantages make the Ritmo an exception among mid-sized compacts. And a model. Time will show that the others will follow." With the Ritmo the buyer, too, gets a jump on the others and becomes a model to be imitated. The Ritmo's technical distinction translates into a social distinction for the owner; the technical improvements in the car promise a social improvement in the owner's play for recognition and prestige: one betters oneself with what is itself better.

The "jump" is the leitmotif, for it creates a consonance between design engineers and car buyers. The striving for this jump contains the common grammatical mechanism that allows the technical sphere to be translated into the social sphere. The jump between the Ritmo and existing automobiles is transformed into a jump between Smith and Jones. That is why each novelty that gains prominence in the market sets the carousel of social competition into movement: it offers the alert buyer a chance to gain distinction over others and leave them behind. Inasmuch as cars are symbols of social superiority, every novelty makes normal automobiles obsolete not only in their technological value, but also symbolically, in that the prestige of an object is rooted in its relative scarcity. Goods available the world over do not create distinction; only scarce goods, those not generally available, are suited for lifting their owners above the masses. The new, which by its very nature cannot (yet) be generally available, clearly has the quality of scarcity; therefore it is especially the new that confers prestige.

The symbolic value of a car is also subject to recall: whoever wants to demonstrate social superiority with an automobile is forced to seek prestige over and over again in new material garb. As a BMW advertisement puts it, "Exclusivity is a malleable concept: from yesterday's to the BMW 300 Series." Precisely those who are ahead have to be careful not to fall behind: "Are you already driving a first-class automobile? Then a test drive in a new BMW 300 can be especially instructive for you. For this new BMW clearly distinguishes itself not only from cheaper vehicles, but unmistakably from comparable ones as well." By continuously producing new innovations, the industrial system is always creating a new generation of scarcities, which in turn set off a new round in the social competition for more valuable symbols. And vice versa: the customer's longing for a jump over the others keeps the demand for novelties alive.

This demand insures first of all that new car buyers do not remain loyal to their acquisition for too long a period of time; after three or at most four years, the new car is dumped onto the used car market. Although neither engine power nor the body has suffered greatly, the symbolic power of a new car has simply worn out. Shiny paint on an old model just does not make an impression. To be able to hold one's own on the novelty market is in itself a mark of social superiority, something people with smaller bank accounts can scarcely manage. More than half of all cars are purchased used; about two-thirds of all transactions occur in the used car market. That means that the novelty policy aims at a rather privileged minority of car owners. And although the policy reinforces their privilege, it is to a substantial extent cofinanced by those who can afford only a used car, for their purchases help the new car owner to move up to an even newer model. Everyone has a car, but not everyone has the same one. Less prosperous groups inherit the faded symbols of yesterday.

The pressure toward higher performance classes also drives the longing for the jump, for the social hierarchy is reflected to a great degree in the size and performance capabilities of automobiles: the higher the car on the product ladder, the higher its owner's place on the social ladder. In 1975, three-fourths of independent and self-employed professionals' cars had a cylinder capacity of more than 1500 cc; only one-third of blue-collar and skilled workers enjoyed equivalent capacity. The horsepower ratios were similar: whereas only one-third of workers had over fifty-five horsepower, two-thirds of professionals did.[4] Social inequalities translate into various engine capacities and vehicle sizes.

Performance is the most important criterion according to which the product palette is laid out; for every bank account there is a performance category. This ordering does not remain stable, however, but is constantly in motion: aspirations for upward mobility push customers to higher performance categories as soon as the pocketbook allows, and on the other side, thanks to the novelty policy, the product ladder

4. *Allensbacher Jahrbuch für Demoskopie* (Allensbach, 1977).

extends ever upward, constantly placing new goals before the eager status seekers. The whole game resembles what happens when one tries to ascend a down-escalator: lots of climbing, but no upward progress. "The technology in the new BMW 300: Yesterday just a wish. Then possible for the very best. Now a model for a new class." Every new round claims to "renew" the automobile and "raise its value." The cycle mobilizes desires for social self-improvement, only to give way to a new indifference because others catch up; then a new symbol of distinction rises on the distant horizon. Novelties keep the carousel of desires for upward mobility in motion, without lessening the distance between top and bottom.

Only the Best Is Enough

Automobiles are a means of communication. Their material characteristics as such do not stimulate demand, so much as the messages that they encapsulate. They are like a language that allows one to enter into relations with oneself and others. They communicate outwardly the owner's self-conception and sense of place in society, and in an internal dialogue they reveal themselves as a source of gloating self-confirmation. The automobile jester H. G. Bentz wears his heart on his sleeve:

> I retreat from the window and slink into the garage, where a certain something sits, gleaming in black enamel and chrome, which means as much to me as a ship to the captain or a thoroughbred to an Arab. I open, an expectant tingling along the spine, the steel-sheathed door and, as usual, stand there for a moment in giddy admiration. Long and broad and low, it lurks like a shell in the muzzle of a cannon. . . . "Well, how're you doing there, Boxie?" I ask and pat it on its broad rump. . . . I pull a gasoline can in front of the grill and take a seat. When I bend down real low like this, the car seems even more powerful, almost frightening, the way it rushes toward me with the crystal eyes of its headlights, the radiator grill, which looks like the windowed facade of a skyscraper, and the broad bumper. From this cozy-creepy meditation I am awakened by a speck of rust I discover on the little horn on the right side of the bumper. Well then, right away I'll have to . . . [5]

Car ownership is not merely a statistical quantity; often it has a narcissistic character. The "glittering something" in the garage: it is mine

5. Bentz, *Alle meine Autos*, 11.

and, what's more, it's perfect. Styling, color, comfort, or technical so-phistication awakens feelings in the owner that, whether of pride or the pleasure of driving, excitement or contentedness, are all alike in that they open areas of self-satisfaction. What one has is what one is worth: this slogan of the commodity-intensive culture, for which hap-piness depends on the consumption of goods and services, has also strongly colored the aura of signification surrounding the automobile. Indeed, the automobile, the most prominent good in the industrial sys-tem, represents the high point of a transformation of sentiments that stretches back a good two hundred years. People have gradually learned to expect satisfaction from things instead of from other people. What was once accomplished privately, from bread baking to house build-ing, or by the local community, from local judicial matters to holiday parades, has been largely replaced by commodities—objects promis-ing satiation and security, entertainment and adventure. The experien-tial space of each individual is furnished with a plenitude of goods; they so stamp our sensations that our feelings become imprints of the commodity world. It is so with the automobile as well: the experiences it calls up become a part of how we experience ourselves, and our de-sire for self-experience directs us in turn to the car.

Novelties thus appear as invitations to self-improvement. Because they rely on the small, innovative difference, they hold out the prospect of new experiences of comfort, speed, or smartness that were not there before, and appeal to the driver's desire for self-development. Who wants to accept the satisfaction of a vehicle from yesterday, when the new model satisfies more completely an even greater number of de-sires? The propaganda surrounding the arrival of novelties amounts to a spectacle of endless change, a world caught up in inexorable strivings for perfection. To become a part of this change promises the customer—however much he or she is otherwise stepped on by the boss or pushed around by a spouse—to change lifestyle a bit, to as-cend to another level of modest happiness, to have another couple of desires satisfied. Since the deep-seated need to transcend daily life is no longer addressed in public festivities or religious practices, the ritual of novelty noisily offers to take over this task, even as it populates small, personal utopias amply with images.

The other side of these messages of fabricated happiness, however, cannot be overlooked: to the advertising industry all people are incom-plete and needy, and it stakes everything on this feeling of deficiency becoming second nature. Products are portrayed as the missing "better half," as an individual's self-completion: "Ideally, a person and an automobile form a unity that functions perfectly, physically and psychologically. . . . It is therefore important for you to find that automobile which best completes your personality." The ad continues, offering the BMW as the completion of a character for which high efficiency is near and dear:

> If you have an interest in modern technological arrangements; if you always like to acquire the most efficient solution, and design quality is an essential element of your expectations of the appliances and tools that surround you, then you should decide on a BMW. With a BMW you achieve a correspondence between your expectations and the performance as well as character of your automobile.

Advertisements from the last few years in particular draw on the most advanced version of pride of ownership—that is, technological narcissism. To admire one's own superiority in the mirror of high-quality technology, to surround oneself with refined electronics or sophisticated driving equipment, and thus to put on airs, is an experience whose form corresponds directly to the model of the efficient, ingenious automobile. The innovations tumble out, one perfection after the other, allowing the quality-coveting buyer to bask in the sun of the latest refinement: technology as a means of gratification. Every improved technology devalues the one immediately preceding it and grants its owner the gratification of having the one that is now most perfect.

Even the sobering side of the automobile, like air pollution and energy waste, has appeared just in time to lend welcome support to the demand for technological solutions. "Porsche makes a lot of steam with a little gas. Why? Because Porsche knows how to solve difficult technical problems perfectly." Here, and especially with Porsche, the issue is not a simple matter of saving gas. Rather, what motivates is the ingenious solution to the efficiency problem, the new degree of technological complexity of which one can take possession. Every innovation stirs impulses to do the decent thing, but also to treat oneself to the newest trick, even if, measured according to use value, it is superfluous. Making technological sophistication—indeed, industrial progress itself, which is after all the project of the society as a whole—one's own plays a big role in the modern self-image. It means being not a mere spectator but an actor in history, and flattering one's sense of self-worth through participation in the state-of-the-art developments of engineering. Overinstrumentalization is seductive. Who has not had the experience of lingering, far longer than necessary, over a really terrific pocket calculator or first-class tape deck. In other words, who has never fallen victim to overinstrumentalization?

Travel and the Tourist's Gaze

"We left Dresden at one o'clock," wrote Otto Julius Bierbaum of the first day of his four-month trip to Sorrento in 1903,

> with Mr. Weber, the owner of the famous hotel, sitting in front as driver; he couldn't be dissuaded from showing us the most beautiful route, since he was at the wheel. . . . We also have Mr. Weber to thank that we took the route through the charming Müglitz Valley, through Dohna, Weesenstein, Glashütte, and Altenberg. That posed a little challenge to our motor, for the road climbs continuously for seven hundred meters. In compensation, the route then falls quite sharply from Zinnwald, the first town in Bohemia, and indeed through a proper old fairy tale forest, in which considerable snow still lay on the ground. We will never forget the journey through this solitude of green and white. It transported us into a world that nearly everywhere is doomed to destruction. Only large estate owners from among the Bohemian wealthy can still afford such forests, in which, so one would like to think, a friendly forest giant rules to whom every tree is sacred and every axe an abomination. In the sharpest of contrasts, just beyond the majestic virgin forests begins the region of the Teplitz coal mines, in which nature has been brutally dispossessed of all its beauty.[1]

In these lines Bierbaum gives voice to a perception that even today is

---

1. Bierbaum, *Eine empfindsame Reise,* 28f.

The landscape as a consumer good. Promenade Automobile, *by A. E. Marty, 1934. From Hervé Poulain,* L'art et l'automobile *(Zug: Les Clefs du Temps, 1973).*

closely associated with the automobile: the automobile opens up the other, contrary world to the daily life of industry. With a car one can let oneself be seduced into distant adventures, homey little towns, enchanting fairy tale forests, far from stale routine, functional ugliness, or the dictates of the clock. The "majestic virgin forests" are uplifting only in contrast to the "Teplitz coal mines"; the automobile offers a chance to flee the latter in disgust and, filled with longing, search out the former. Here we find a paradox, yet one that is common in the history of the technological mentality: the most modern creation of industry as the bearer of anti-industrial impulses.

The Consumption of Landscapes

One finds them in parents' and grandparents' photo albums: pictures, in dull black and white, of a car against a romantic backdrop, with (usually) a female leaning with stiffened grace on the hood in the foreground—the glaciers of Grossglockner or the city walls of Rothenburg bear witness to a delightful automobile tour. The automobile periodicals of old are also full of route descriptions and travel reports, of adventurous and pious descriptions of the joys of tourism by car. They indicate how much the automobile was perceived as a means of enter-

"See the world in your Mercedes-Benz! . . . Mercedes-Benz—The car in a class by itself." Advertisement for the International Automobile Exhibition, from Die Dame, *December 1928.*

tainment and how little anyone, for half a century anyway, saw it as a vehicle for simple commuting. Experiencing the world as a tourist first became popular with the automobile. The automobile not only opened new arenas of self-experience, but it cast space and geography in a new light as well.

The tourist's gaze examines the world in terms of three motifs: "untouched nature," "unspoiled customs," and "incomparable sights." Cliffs and ice, once considered threatening, become welcome sights; the unadulterated life appears in processions and native dress that otherwise are seen as backward; city towers and quaint gabled houses, subject to demolition according to need, come to represent a devotional testimony to a surprising past. As if paging through a picture book, the tourist now travels through the land, at once touched and amused, holding watch over the peculiar, the unusual, the unique. "Above all, however," one traveler swooned in 1929 about his spring tour,

> [is] the poetry of the journey, the fairy tale world of streets and distant locales and their miracles and adventures. . . . The idylls into which one is suddenly transported: a quiet country path over whose banks corn flowers and red poppies send their greetings from rolling fields, playing children in twisting village lanes, peaks rising from distant mountain ranges, sunset in the low plains of the flatland, forests with sunlight dancing through the leaves confusing the senses, moonlight on mountain passes, blonde girls and their welcoming waves. . . . And then one more thing: one need not drink the landscape in great draughts, but here and there as well with the little sips of an epicure abiding in leisure.[2]

Pretty much all the clichés of a romantic view of nature are gathered here. The land becomes an idyll, nature a fairy tale world—and the gasoline vehicle, that rattler and stinker, is the fairy that abducts you there. Thus is the automobile bound up with an attitude in which nature and people are the raw material for sentimental devotions, where changes in the landscape, like changes in the lifestyle of the local inhabitants, become a means of gratification. "To drink the landscape here and there with the little sips of an epicure": the language betrays how the traveler perceives what lies along his way. Under the tourist's gaze, the world becomes a consumption item. Clearly, this perception can develop only from a distance; indeed, it stands opposed to the perception of the local inhabitants. Those who make their living fishing on the Côte d'Azur or producing firewood in the Black Forest pay to the sea or the forest a wholly different kind of attention, one having little in common with what the traveler sees. Only the outsider, the stranger who has nothing at stake, can afford the tourist's gaze; it is a perception that lives off of distance. For the tourist, even the dock-worker's drudgery can be an aesthetic spectacle:

> Whoever wants to see the Hamburg harbor in its Sunday dress has to see it on a sunny workday. I know of no more powerful image of work. It is as if all the sounds of the world have combined here into a ringing, rolling, buzzing, hammering, grinding, whistling, clattering, howling, sighing, thundering symphony of work.[3]

Only one who does not have to toil there, who is no worker, speaks in this way, just as those who "drink the landscape" do not have to plow.

2. *Der Motor-Tourist*, 1929, no. 9, 8.
3. *Der Automobilfahrer*, 1931, no. 5, 67f.

On the contrary: the attraction a place has for the tourist grows in direct proportion to its distance from the conditions of the tourist's own life. The contrast with daily life allows romantic feelings to sprout.

A stream of motifs stretching back to the beginning of the nineteenth century—for England, into the eighteenth—find renewed expression in the experience of the automobile trip: acquaintance with nature as the devotional opposite to the cold and inexorable regularity of industrial society. The "landscape" is the construct of a society that no longer has an unmediated relationship with the soil; "peasant customs" are romantically honored in places where the cities regard themselves as the producers of wealth; and even the "sights" take on an aura of awe only once their "capriciousness" has been annihilated by city planners' notions of order. It is not by accident that the nineteenth century saw the rise of landscape painting, in which dear nature is represented completely devoid of people and apparently untouched by their intervention. Such paintings celebrated the natural field in its primitiveness and harmony, and swooned over the solitary bliss of the forest and walks in the woods, which promised an unalienated feeling of life. Nature was discovered as the ideal counterpart to the world of calculated bustle that had begun to spread among the gray walls of the city. And so the first trips for the mere sake of enjoying the landscape began: the melancholy highlands of Scotland, the blue grottoes of Capri, the Rhine Valley with its cliffs and castles and the enchanting song of Lorelei—all these became classical sites of devotion. Professional assistance was also part of the plan: in 1839 the Koblenz bookseller Karl Baedeker published his first travel guide (about the Rhine), and in 1841 the hiking enthusiast Thomas Cook founded in London the first travel agency.

Thomas Cook organized tours with the railway, and it was also with the railway that the tourist's gaze, geared to the extensive consumption of landscape images, achieved such enormous popularity. Seen through a train window, the landscape offered itself up like a panorama of ever-changing scenes, while the foreground rushed by in a blur.[4] Sequestered inside the compartment, one could smell and hear nothing from the world outside; thus travelers' projections, more so than their own experience, determined how the landscape should be perceived. Technology made even wild regions accessible, yet at the

4. Wolfgang Schivelbusch describes the new mode of perception introduced by experiencing speed on a train, in *The Railway Journey: The Industrialization of Time and Space in the Nineteenth Century* (Berkeley: University of California Press, 1986).

same time it created distance; because of this distance, unpeopled moorish heights could be experienced as painterly, and a yawning precipice as suspenseful. The railway choreographed a landscape that then took on a monumental dignity.

This perceptual legacy also marked the experience of traveling by automobile, but with a twist: the automobile broke the rigid perspective as seen through the train window because, freed from the tracks, it could change direction and speed at the driver's will. The landscape became accessible from all angles, so that fixed perspectives dissolved into an abundance of views, a multitude of vantage points. Moreover, the automobile transformed the traveler into a potential explorer, who could set new goals and approach the old in a fresh manner; now, with active probing, the tourist's gaze could penetrate even the most distant corners.

At the turn of the twentieth century, following the massive and violent surge of urbanization, the impulse to go into the wild and green became a much-discussed topic. Insofar as the romantic gaze sought out experience contrary to that of progress in the cities, it was precisely there, amid the noise, the crowding, the poverty, and the social turmoil, that the need arose among the bourgeois to get away from it all and go in search of blue blossoms. The automobile inherited these desires and lent them a new expression. Aside from allowing city dwellers to travel, however, it also made possible a response to the traditional tension between the city and nature nearly unknown until now: the suburban lifestyle. Bierbaum already sensed something far-reaching to come: "With the automobile it becomes possible to live a couple of hours by train from the city and still go to the city to pursue one's employment."[5] The rush to the suburbs aimed at fresh air and trees in bloom; it was and is an attempt to seize space so as to enjoy the advantages of the city and proximity to nature at once, without being touched by the disadvantages of either. Especially in the United States the automobile seemed to offer a way out of the urban crisis, to be a means of "ecologically" reforming cities, which were suffocating in industrial noise, horses, and new rural immigrants, and effecting a reconciliation between city and country. "Imagine," an American author wrote in 1904,

> a healthier race of workingmen . . . who, in the late afternoon, glide away in their own comfortable vehicles to their little farms or houses

5. Bierbaum, *Eine empfindsame Reise*, 281.

in the country or by the sea twenty or thirty miles distant! They will be healthier, happier, more intelligent and self-respecting citizens because of the chance to live among the meadows and flowers of the country instead of in crowded city streets.[6]

The city was to be cured, according to Henry Ford, by leaving it. Not that this remark was surprising, given the special sort of pollution that stirred the senses: the horses in New York City alone deposited 2.5 million pounds of solid waste on the streets every day, and as much as 60,000 gallons of urine; and 15,000 dead horses had to be removed from city streets every year.[7]

Recreation on Wheels

Vacations and cars—how many people fail to think of sand and sun, distant places and unfamiliar people, when they buy a car? The vacation trip is the quintessential focus of anticipation, the stuff of which automobile dreams are made:

> Once again our little Mathis sits in front of the garage, packed full with everything needed for a big trip. There's nothing superfluous, and still there's barely room in the trunk for a thermos bottle and some provisions. A roomy camping tent, including blankets, lies sideways across the car's canvas top. We check the list carefully to make sure everything is there, everything is in order. . . . Tomorrow we take off! Once again tonight we'll dream the dream of the past months: blue sky, blue sea, tall palms, slender pines, rugged cliffs, strange houses, unfamiliar people. The French and Italian Riviera is our destination for the coming weeks.[8]

This mood of an imminent departure as described in 1929, and since then longed for and experienced by millions, is indulged to the fullest in many fantasies about the automobile. It combines a longed-for freedom from rules and routines with the pleasure of self-determined mobility; in both, the wish for an unregimented life is active, the common root that holds them together. And so once again the popular hope for escape from the iron cage of industry seizes hold of the automobile: the longing for another time and another world, for a break from

6. Quoted in James Flink, *The Car Culture* (Cambridge, Mass.: MIT Press, 1976), 39ff.
7. Ibid., 34.
8. *Der Motor-Tourist,* 1929, no. 13, 4.

Cover of Elegante Welt, *1926, no. 13.*

work and the bustle of the big city, has become a twin to the automobile. The automobile is, in this sense, a deeply romantic vehicle.

The vacation trip and the automobile, moreover, found their way into society and became items of mass consumption roughly contemporaneously. Until the beginning of the nineteenth century there were three distinct conceptions of travel: the European "Grand Tour," undertaken especially by the young, a marshal's baton in the rucksack, for purposes of self-education; the trip to Italy for an encounter with antiquity, a station along the road of self-discovery; and the trip around the world, a journey back to the wild naturalness of the original human condition. Soon, however, aristocrats and wealthy bourgeois (especially among the English) began enjoying the sea, air, and sun, taking off on trips to the baths and the ocean.

They went first for the sake of relief from their maladies, then for

We Germans have become leaders in the automotive industry; we also gave the significance of reliability trials its rightful due. In short, we have taken a big, perhaps the biggest, part in the victorious and portentous development of the automobile. But certain things involved in imparting to it the art of living we have heedlessly passed by. That is simply in our nature as a working people. Just as the French are always gratifying their effervescent, explosive temperaments in sensational high-speed races, so will we always remain in principle a basic, goal-oriented nation. The English, in contrast, are not unfortunate in their ability to lighten some of the strictness of sporting activities through a genial, almost quaint sense of humor, through a joy in calm, agreeable living. We Germans immediately recognize this English talent, the more so whenever an Englishman has sought out such hours of friendly existence in nature. For that reason, our healthy capacity to conform—which, as is well known, we possess more than any other nation—might bestow upon us a side of motorism that is attractive and worthy of imitation: "camping." The English understand "camping" as a free and easy style of pitching tents in the wild; the experience stretches over days, often even weeks, and is an original connection between culture and nature.

Hidden deep in the silent landscape, in a forest meadow full of tall grass, near a gentle little stream, sits the motor car. A rainproof awning is stretched over it. Beside it are the "wigwams," likewise made of a friendly light grey or white cloth. A wood fire flickers under a humming blue enamel kettle being tended by a lady's delicate hand. In her city apartment she would not make the tea. Here, however, she does; for one does not take the domestics along on a "camping" trip. It is just as obvious that they are left at home as that the otherwise customary and

recreation and leisure. Nice, for example, was at the beginning of the nineteenth century a winter refuge for ailing English aristocrats; later it served as an extravagant winter lodging for European elites, only to develop in the 1930s into a summer vacation paradise for the middle class; and now, of course, it is the site of mass tourism. The decisive transformation was from the leisure classes, seeking a mild winter in an exclusive atmosphere, to the employed classes, who sought an alternative to the world of wages for two unclothed weeks of recreation in the sun. Vacation, in the sense of a paid period of rest and relaxation, was, after all, entirely unknown in Germany before 1875; only in that year were employees of the Imperial Post Office granted eight to ten valuable jewelry is deposited in the bank vault—unless, of course, one is traveling in refined company. White and airy, too, is the thoroughly informal attire; one might even go barefoot or wear straw slippers. The children leap about in the freest style conceivable, their joy fluttering in friendly competition with the loose hair of the girls in the breeze.

The "chief" of the camp is just approaching. His sleeves are rolled up, revealing two suntanned forearms; in one hand he carries a fishing rod, in the other a small collapsible aluminum bucket, in which the patiently angled dinner (fried fish is on the menu) flops. He—one of the most significant figures in the upper elite—carries his sustenance with pride, the fruit of his own labor. In just a few minutes the folding furniture is removed from the bottom of a large suitcase and set up; a little table cloth is spread over it, and now, from deep within the attractive picnic basket, which holds all of the delicious, sweet secrets and found a convenient spot beside the

spare tire on the drive out, many delightful treats emerge: the cookie tin, the sugar bowl, a little bottle of rum. . . .

No, one truly misses nothing in this corner of the world, completely isolated though it is, because such a significant means of assistance—as the motor car—belongs to it.

Allgemeine Automobil-Zeitung, April 1914.

From Elegante Welt, *1926, no. 16.*

days' leave, and as a consequence the practice devolved to the benefit of all civil servants and employees in trade and industry as a means of securing their loyalty. Workers had to wait longer; aside from the occasion proof of goodwill from a socially minded entrepreneur, they knew nothing of vacations until the twenties and thirties. Throughout these decades, the desire for travel and recreation grew and became bound up with aspirations for a car, part of the motorized conquest of space as an arena of recreation—promised, incidentally, in the "Power Through Joy" propaganda of the National Socialists and ultimately realized in the economic miracle of postwar Germany.

Instead of the glow of sunset on the Swiss Alps or roulette on the

Riviera, the leisure desires of the less well heeled were focused on out-ings or on Sunday drives. Here, too, a new world grew up around the automobile:

> To the automobile owner a weekend in the car meant relaxation, care for the nerves, the chance to participate in water sports, to socialize in nature free from convention. . . . There is simply nothing nicer than an aimless spring or summer drive under the blue sky through fields and forests, over hill and dale. One races to where nature beckons, and then drives on unfettered, without plan, wherever one's inclinations suggest. Thus does the driver become acquainted with all the beauties of the homeland and learn to love them, as was not possible in earlier times.[9]

Around the automobile, too, grew new standards as to what is beautiful and important and worthy of effort in life—a construction of reality, so to speak, that casts nature as well as space in a new light and allows experiences and pleasures scarcely known before. Whole regions were reconceived as recreation landscapes (and then recon-structed), offering people a diverse backdrop for their outings. Appro-priate practices and models of behavior—the aimless drive under blue skies, ski trips, camping—came into fashion.

> In many places in the homeland, particularly on lakes with wooded shores, improvised car and camping cities populated by hundreds of cheerful automobile travelers spring up on the weekends. A unique, alluring life of high primitiveness and wholly unforced freedom from convention develops in the open air or the tent. People cook and sleep, eat and play, dance to music on a gramophone or the radio, paddle in their portable boats, row, and partake in other sports.

Thus does a new rhythm of time, a new lifestyle, crystallize around the automobile. What one thinks of the landscape, what seems important on the weekend, what experiences prove restorative—the world is set right from behind the wheel of an automobile.

9. *Die Auto-Schau. Das illustrierte Jahrbuch für Freunde und Käufer des Automo-bils* (The car show: The illustrated yearbook for friends and buyers of the automobile) (Berlin, 1934–1935), 95.

Space and Time as Resources

On the occasion of the inauguration of a motorized postal service between Friedberg and Bad Nauheim, the *Allgemeine Automobil-Zeitung* considered the difference between the traditional means of locomotion, the team of horses, and the new one, the motor wagon. It culminated as follows:

> Those who earlier wanted to view the beauties of the German fatherland had to have much time and took much time to arrive at their destinations by the venerable old means of transportation. Today the pulse of life beats more rapidly, and traffic steadily increases. One wants to be conveyed as far as possible in the shortest possible time, and so our fast-paced generation has created for itself a new means of transportation. Everywhere new networks of motorcar traffic are arising, replacing the old motorbus and postal routes.[1]

An odd astonishment is sounded in these statements. The memory of more prudent times is still alive, but contemporaries seem fixated on a better idea: "to be conveyed as far as possible in the shortest possible time." Rushing and nervousness, haste and impatience flicker through the urban dweller's feeling for life, as in film (cinemas were just now beginning to spread) one event followed on the heels of another, storming by. Indeed, geography itself appeared to totter: what

1. *Allgemeine Automobil-Zeitung,* 1906, no. 31.

was happening in New York seemed no less significant than what transpired before the front door. Telegraph and telephone, rotary presses, and a flood of photographs delivered even the distant world to the eyes and ears of everyone, parading events in quick succession. Daily life around the turn of the century was dominated increasingly by a shortage of time and a love of distance; one gazed up at the zeppelin and quickened the pace. After a lengthy labor, the "fast-paced generation" was finally born out of the nineteenth century. Although obstinately greedy with minutes, they were generous with kilometers; and it was in the automobile that their attitude toward the world found material expression.

As the train station clock was to the railway, so was the wristwatch to the automobile. During the same decades that the automobile made a place for itself in society, the wristwatch became established, making the deliberate glance at the wrist the most pointed gesture of the industrial age. The wristwatch descended from the factory and train station clock, both of which stand for the temporal regime that had spread over society during the nineteenth century. With the factory clock workers in the plants were compelled to follow the relentless beat of minutes and hours, a regime designed to rid them, for the greater good of commodity production, of all idleness and self-determination. Because it controlled both the beginning and the end of the workday, moreover, the factory clock extended its dominance all the way into the family, thus subordinating the world beyond the factory to its regime as well. With the train station clock, the cult of punctuality overtook the whole of society, for after 1850 the time atop the train station became the standard. Whereas formerly people had followed the local course of the sun, now the same time became obligatory for all.

It is no accident that the great European train stations were constructed around the clock; with their portals, domes, and massive halls, they often seemed like cathedrals dedicated to the precisely timed comings and goings of masses of people. The clocks that urged workers and passengers to hurry were still mounted for public view, but with the advent of the pocket watch individuals began to carry around admonitory pointers of their own. The rational treatment of minutes and hours thus became a matter of continual self-control, and soon no niche in daily life was immune: the kitchen clock presided over eating times, and with the alarm clock punctuality took over the bed. The wristwatch did away with the leisurely reach into the vest pocket, so that from now on, with a deliberate turn of the wrist, every impulse could be checked for whether there was time enough to accommodate it. The admonitory pointer became omnipresent and was internalized in emotional life—whatever one did was overseen by a glance at the watch. In a time of universal respect for time, it was only logical that the automobile would be welcome, as one said, as a "time-saving machine."

Accelerated Sales

The tourist's gaze is to the trip-happy motorist as the efficiency expert's gaze is to the sales-happy businessman. Every commodity, once produced, presses forward to be sold; net proceeds are not to be had otherwise, which return the capital invested, padded as much as possible with profit, to the coffers of the enterprise. Little but rust and mildew can live from goods stored away in warehouses; only when their sale proceeds smoothly do they fulfill their function and keep the wheel of production turning. Productivity gains in the factory are realized only when sales increase, that is, when the market expands geographically and the goods find their way into the buyers' hands more quickly. Economic growth is nourished by accelerated turnover just as much as from increased productivity; that is why the economic gaze is unremitting in its efforts to distribute goods and services both more broadly over space and more densely within it, thus to achieve a more rapid turnover in time. As long as space and time represent immovable obstacles, as long as distances cost a fixed amount of time and a unit of time suffices only for a fixed distance, then turnover and growth are braked. It was only because ships, railways, and then the automobile mobilized (and so reduced and saved) time and space that growth could pick up speed. For the economic gaze, space and time are thinned into mere resources, which can be compressed and used more efficiently through a particular means of transportation to increase the turnover of goods and services. This endeavor met the automobile halfway and conducted it to new heights: "Among business travelers," a 1923 commentary states,

> another series of arguments weighs heavily in favor of the automobile. There is, for example, the possibility . . . of being able to find the customer, if he is unable to meet at the moment, at another suitable time of the day, if necessary, assuming he is in a neighboring town or, as is often the case for us, working in the field, seeking him out even in the field. . . . Above all, however—and this is the decisive point—there is the possibility of cultivating the individual customer much more inten-

sively, with much shorter intervals between visits, without having to increase the traveling staff. These, summarized in a few words, are the advantages of traveling by automobile.[2]

Space and time no longer present boundaries to which one must adapt, but have become manipulable, exhaustible resources, efficient use of which can refine and broaden market accessibility to increase retail turnover.

To professional groups, such as buyers, farmers, and rural doctors, that operate within moderate distances and so are poorly served by the railway, the automobile was particularly important for the sake of rational use of space and time. After the railway established the slogan "time is money" as fact in matters of long-distance travel, regional transportation also accelerated; here, too, the automobile became the key to the comprehensive capture of regional markets. Doctors and landowners, fire departments and ambulance services all saw in the automobile an instrument for accumulating time in order to turn it into greater distances. Overcoming the power of space to separate was particularly imperative for those who wanted to stay ahead in their competition with other suppliers. The battle for market power was founded— for doctors as much as for fruit sellers—on acceleration. The more the economic gaze grew in power, the more it supplanted all other modes of perception, thus determining the transformation of society, and the more traffic policy became concentrated on tailoring cities, villages, and the landscape itself to the rapid circulation of goods. Competitive advantages, after all, required a space ever more hospitable to passage. A discourse grew in which (as was soon evident in all political debate) everything turned on "bottlenecks": because they raised the specter of stagnant circulation, they had to be fought with more transit connections and faster means of transport; otherwise, it was said, growth would collapse. Nevertheless, this discourse led to a dead end: the pressure of competition only created ever more circulatory bottlenecks on a new scale. To overcome them proved an endless endeavor, and finally nothing was left but perforated space.

The Compression of Daily Life

While the automobile gave producers and dealers the chance to eat up minutes and kilometers through accelerated circulation, it provided

2. *Der Motor-Tourist,* 1929, no. 13, 4.

consumers the chance to connect with the circulation and gain access to more goods and services. With the automobile comes a linkage to the colorful world of urban consumption; with the automobile the rhythm of daily transactions increases in density. The shortage of time for consumers in an advanced commodity society results ultimately from the increase in offerings, whether types of cheese or cinematic films, that make demands on their time, which is itself not subject to increase.

How such an increase in density creates a new need for planning and rationalization in daily life is nicely illustrated in the following recommendations from 1931:

> It is much more necessary to plan an automobile trip conscientiously when one wants to get from it everything it offers in the way of beauties. . . . One therefore goes over the intended route very precisely on a map, with the help of travel guides, and plans sufficient time to view these things. . . . The experienced automobile traveler who wants to visit and become acquainted with distant areas does not lose precious time visiting nearer places. For once underway it is often difficult, indeed nearly impossible, to determine at every instant what there is to see and how far one must drive to reach a particular spot. It is therefore highly recommended that the most important sights be briefly noted on special cards for each day with catchwords from the travel guides.[3]

For the last hundred years or so, initially among the higher social strata, the complaint that time rushes by has permeated the experience of daily life. The hours of the day seem insufficient to take care of all our desires, intentions, and demands. Generally, the feeling of a lack of time begins when so many attractions and obligations press that they threaten to exceed the available time. Expectations then outpace capacities, and the discrepancy between what one should do and what one can do becomes chronic. With the increasing plenitude of experiential possibilities, with the rising number of goods and services that promise happiness, one begins to feel that more and more is needed and that one can never acquire or accomplish enough. The fleeing of time, in short, arises from a chronic "deficit consciousness."

Thus we find reflected in the microcosm of daily life what the nineteenth century postulated for the macrocosm of history: the future would endlessly surpass the present, with the progressive production of goods and services satisfying the endless needs of people and leading

3. *Der Automobilfahrer*, 1931, no. 10, 148.

On this first occasion I was unable to go to Raspelière alone as I did on other days, while Albertine painted; she wanted to come there with me. Although she realised that it would be possible to stop here and there on our way, she could not believe that we could start by going to Saint-Jean-de-la-Haise, that is to say in another direction, and then make an excursion which seemed to be reserved for a different day. She learned on the contrary from the driver that nothing could be easier than to go to Saint-Jean, which he could do in twenty minutes, and that we might stay there if we chose for hours, or go on much further, for from Quetteholme to la Raspelière would not take more than thirty-five minutes. We realised this as soon as the vehicle, starting off, covered in one bound twenty paces of an excellent horse. Distances are only the relation of space to time and vary with it. We express the difficulty that we have in getting to a place in a system of miles or kilometres which becomes false as soon as that difficulty decreases. Art is modified by it also, since a village which seemed to be in a different world from some other village becomes its neighbour in a landscape whose dimensions are altered. In any case, to learn that there may perhaps exist a universe in which two and two make five and a straight line is not the shortest distance between two points would have astonished Albertine far less than to hear the driver say that it was easy to go in a single afternoon to Saint-Jean and la Raspelière. Douville and Quetteholme, Saint-Mars-le-Vieux and Saint-Mars-le-Vêtu, Gourville and Balbec-le-Vieux, Tourville and Féterne, prisoners hitherto as hermetically confined in the cells of distinct days as long ago were Méséglise and Guermantes, upon which the same eyes could not gaze in the course of a single afternoon, delivered now by the giant with the seven-league boots, clustered around our tea-time with their towers and steeples and their old gardens which the neighbouring wood sprang back to reveal.

Marcel Proust, *Remembrance of Things Past*, trans.
C. K. Scott Moncrieff and Terence Kilmartin, vol. 2:
Cities of the Plain, chap. 3 (New York:
Random House, 1981), 1029.

them on toward ever greater happiness. This conception of society characterizes people as perpetually needy beings, who depend for their contentment on the delivery of ever more clever products of all kinds. As soon, however, as the notion of unlimited desire as both morally in order and rationally necessary gained currency—and this marked a decisive break in the history of virtue during the eighteenth century—then a conflict was destined to emerge: unlimited desire ran up against limited time. The day, in its conservative manner, remained in the end only twenty-four hours long despite all the progress, and the clocks, too, ran exactly as fast—or, more to the point, exactly as slowly—as they had for centuries. So, because the number of hours could not be increased, the only choice was to get more out of them, through either haste or planning.

The more opportunities there were for new experiences, the more the offerings competing for the limited number of hours enticed, the more valuable time became: an expensive commodity that had to be saved, exploited, and carefully planned. Soon it was a sacrilege to waste time in the present, since tomorrow always has a stronger attraction than today, and the successful life was everywhere a task of the future. Given such an emotional orientation toward life, the acceleration of all processes took on the air of being all but a moral obligation, pushing the drive to happiness into a continual race with passing time.

It is part of the logic of this worldview that space and time are experienced as enemies of happiness, as the basic forms of hindrance, which must be opposed with all available means.[4] Every distance interposed between need and gratification becomes an irritation; every second of waiting, which only pushes gratification into the future, is a loss. To obliterate distances in kilometers and hours—indeed, to do away with space and time—that is the hidden utopia of a society crazed with the future. When success is always lacking because new promises continually appear on the horizon of the future, the passing of time and the limits of space become threatening on an existential plane. What else, then, is to be done but to make every effort to approximate simultaneity and omnipresence more nearly? The praise that greeted the automobile in 1906 cannot be understood otherwise:

> The automobile has succeeded in overcoming space and time more completely than ever before. From a certain point of view, all of the technological endeavors of mankind are directed only at transcending space and time. The telegraph transmits the written word over great distances in the smallest time units, just as the telephone does for the spoken word. . . . Steam locomotives and electric trains, steamships, bicycles, airships—all of these serve the purpose of conquering space and conquering time, to race across the greatest possible distances in the shortest possible time.[5]

In the battle against expanse and duration for the sake of satisfied desires lies the metaphysical meaning of the transportation revolution, as a technological offensive against the transient nature of life.

4. Günter Anders explores how time and space become basic forms of hindrance when needs are considered unlimited, in *Die Antiquiertheit des Menschen* (Munich: Beck, 1980), 2:335–41.

5. *Allgemeine Automobil-Zeitung*, 1906, no. 17, 33.

*Peugeot accélération, by Paul Colin, 1935. From
Hervé Poulain,* L'art et l'automobile *(Zug: Les Clefs
du Temps, 1973).*

This passion to reduce distances to near zero in order to retain time
at its fullest certainly stimulated the imagination of the engineers. Consider the following futuristic utopia, which speculates from the perspective of 1909 about transportation technology in the year 1965:

> My American friends . . . energetically insisted that I help them in the
> construction of a tunnel railway for travel from New York to Paris at
> the speed of sound. . . . The old pneumatic mail principle was to be
> used. The tunnels were fabricated into precise cylinders and the trains
> outfitted front and rear with seals that fit the cylindrical tunnels like
> pistons in an engine. . . . At the end of the tunnel, in front of the train,
> gigantic air pumps sucked air out at a hundred thousand cubic meters
> per second to create a fairly complete vacuum in the tunnel during the
> journey, while at the other end pressurized air streamed out of enor-
> mous containers to push the train forward. By 1965 the railway was

finished, and we made the first trip from Paris to New York. The tunnel was exactly seven thousand kilometers long, and, according to calculations, we were to cover the distance in five hours and fifty minutes. . . . Before we descended into the tunnel in Paris, we lingered for a short time on the bottom level of the Eiffel Tower and watched as the sun slowly sank in the west, how the reddish gold disk touched the horizon and disappeared behind it. At the moment it was gone, my American partner grabbed me by the arm and cried, "Quick, down into the train, we'll catch up with it again in New York."[6]

6. *Das Neue Universum* 30 (1909): 4f.

Disenchantment

Photograph by Robert Rodale.

The Aging of Desire

Running out of gas, Rabbit Angstrom thinks as he stands behind the summer-dusty windows of the Springer Motors display room watching the traffic go by on Route 111, traffic somehow thin and scared compared to what it used to be. The fucking world is running out of gas. But they won't catch him, not yet, because there isn't a piece of junk on the road gets any better mileage than his Toyotas, with lower service costs. Read *Consumer Reports,* April issue. That's all he has to tell the people when they come in. And come in they do, the people out there are getting frantic, they know the great American ride is ending.[1]

With these lines John Updike opens *Rabbit Is Rich,* his novel about the splendor and misery of the 1970s, expressing the sentiment that underlies the gin-drinking, numbingly self-satisfied world of Rabbit Angstrom in the petty-bourgeois town of Brewer: the great American drive is coming to an end. The excitement over the automobile has gone flat, a tiredness is setting in, and conflicts are arising in the political arena. In the 1970s there grew for the first time a discontent with the automobile: fathers had to make household budget cuts to be able to pay the higher price of gasoline, and politicians suddenly had to heed countless civil protests and curry a distanced relation to the auto-

---

1. John Updike, *Rabbit Is Rich* (New York: Alfred A. Knopf, 1981), 3.

mobile in order to maintain their popularity. The days of great promise were over, the times of heady planning past; disillusion pressed to the fore, and cynicism even broke through.

What had happened to the high-flying project of motorization? Where are all the expectations and desires—the hope, ultimately, for a better life? It is not as if they had disappeared: 1.2 million visitors attended the 1983 International Automobile Show to admire the newest creations. But the consensus of the early decades was broken: while some still got a shine in their eyes, others laughed scornfully as they danced around an idol of the golden calf, which was erected by protesting citizens' groups outside the show. The great transformation of the 1950s, accompanied by the drumroll of progress and prosperity, had come a long way and unquestionably changed the face of Germany profoundly. But it had led to something completely different from what our fathers had hoped. The air has leaked out, the dreams have turned stale, and even the hymn of praise of the unwearied has the sound of a spiteful "nevertheless": the love of the automobile has cooled.

The historical project of motorization—pressed forward economically since the beginning of the century by industry, incorporated socially through the striving of social classes for prestige, and colored culturally by exciting images and dreamworlds —lost its power to persuade at the very moment of its triumph. Indeed, the failure was so profound that some wanted to turn back the wheel of history, if they only could, among them not only citizens demonstrating against the highways eating up the landscape, but ministers in the seats of power as well. A change of theme had occurred: whereas earlier the drives and ambitions of entire social epochs had reverberated in the automobile, the consonance was broken in the seventies, its tones turned shrill. The automobile itself had not changed, but the passions and utopias embodied in it had lost their buoyancy. New images of the good life were gaining ground, for which the automobile (and with it the ideal of a society made for cars) had become quite nearly the essence of modernized wretchedness. The promises of motorization, it appears, have left behind automobiles without promise.

When a federal minister proclaims the automobile "enemy number one of the environment," when street construction projects provoke demonstrations, when the bicycle enjoys a new popularity, then clearly a break has occurred in the history of automotive enthusiasm. Desires and objects are not married eternally to each other; they can fall into conflicts and drift apart, until finally the relationship is destroyed. Desires make their retreat when, over the long run, they are disappointed and subverted by contrary experiences.

They can search for other objects offered by technological progress—the current enthusiasm for microcomputers, for example, reflects a leitmotif or two from the early years of the automobile. Otherwise the desires simply get drawn into a whirlpool of doubt, losing so much legitimacy that they become obscured by other hopes and values. Just as the needs expressed in the automobile were, generally speaking, born in the second half of the nineteenth century and represented a break in the history of needs at that time, so can they age and pass away at the end of the twentieth: in both instances a turning point in the history of culture is marked. Needs are not facts of nature. They are learned and can be unlearned when circumstances are no longer hospitable to them.

From Traffic Jam to Traffic Jam

> Summertime is surely the nicest time of the year, since everyone has time. And when the sun shines, people want nothing so much as to get away: quick, quick, to the sea or the mountains. Look out, the time for vacation has come! Between the wish and reality runs the Autobahn, a gray band 7,919 kilometers long dividing travelers from their destination even as it links them to it. And on weekends between June and September in Germany all hell breaks loose on the highways, and every year it is the same. Vacationers rush off en masse at the same time in the same direction, all wanting if possible to arrive on the same day they depart. Then everything races to a halt in a traffic jam—to everyone's great dismay . . . a snail's pace, delays, accidents—in temperatures nearing the nineties. A police spokesperson called the Autobahn "the longest sauna in the world."[2]

What most damages the automobile's attractiveness is its success. Mass motorization itself is responsible for bringing experiences in tow that undermine enthusiasm for the automobile. This fascination was, simply put, born in a time when the automobile still possessed the value of a rarity, in that only a few people possessed one. The feeling of being independent of the masses and not ruled by schedules flourished only so long as the streets remained free of auto avalanches and traffic regulations, just as the sportive joy of speed first gained its power of

2. *Die Zeit*, 1983, no. 32.

attraction when free roadways yet beckoned and one could race triumphantly past carts, carriages, and the early midget motorcars. The joy of driving, in short, rests largely on relative advantages—that is, on advantages that others do not enjoy, because they do not possess an automobile and are therefore relatively immobile and slow.

With mass motorization, however, the picture has changed, and the relative advantages the automobile once conferred have dwindled: the more cars, the less joy. Now the masters of space and time are held captive by clogged streets, now the pleasure of speed falls by the wayside of full highways and high-powered competitors. The desires become fragile because the conditions under which they first grew up no longer pertain. Driving is no longer reserved to the few able to enjoy their privilege at others' expense; on the contrary, many have in the meantime squeezed behind the wheel, and now their cars hinder other cars, their desires get in the way of other desires. Privilege evaporates in this mutual blockade, with disillusion spreading in its place. The desires get old because experience continually denies them. Modern street traffic destroys the very hopes that created it.

It is apparent in hindsight that the utopia of mass motorization rested on the illusion that the pleasure of early motorists could add up to a general mobility prosperity for the masses. Yet this utopian projection failed to consider that the desires of individuals—in a space subject to limited enlargement—will necessarily run into and diminish each other, and as a result the anticipated prosperity lagged far behind expectations. The automobile belongs to a class of commodities that cannot be multiplied at will. Because its attraction requires the exclusion of the masses, the democratization of car ownership destroys its advantages.

The more drivers there are populating the streets, the more the desired jump in speed shrinks; but in the situation of a general traffic jam—on late afternoons in the city or Sunday evenings on the highways—not only do relative advantages shrink, but new burdens grow up as well. Once a certain traffic density is surpassed, every approaching driver contributes involuntarily to a slowing of traffic: the time that the individual driver steals from all the others by slowing them down is greater many times over than the time he or she might have hoped to gain by taking the car. From traffic jam to traffic jam, the time needed to get from the office to home or from home to the vacation spot gets longer; with every additional car each minute is worth less in distance. Tem-

From Hervé Poulain, L'art et l'automobile *(Zug: Les Clefs du Temps, 1973).*

pers flare; everyone is stealing everyone else's precious time; annoyance reigns everywhere, and rage: the pleasure in time won turns into fury over time stolen. And in view of the clogged streets, there rises in the throat a cry that, again and again in the history of motorization, has swelled to a shrill chorus from all the frustrated drivers: we need more streets and we need them now!

Under the title "The Metropolitan Traffic Crisis" (subtitled "Permanent Congestion in the Streets—Pedestrians Faster Than Cars"), the *Berliner Illustrirte Zeitung* had sounded that call as early as 1926:

> Not only in all-too-crowded New York, but even in spacious Berlin, there is tedious congestion at certain times and places, for example, Brandenburg Gate. . . . On the same number of city streets as before the war, three, four, five, and six times as many vehicles, most of them motor vehicles, are driving now. . . . The question involves increasing the number of traffic arteries; it is a problem for the city planner, not

the police. The city planner must, finally, construct the necessary tunnel and bridge passages—above and under the ground; that is the only solution. It must be done quickly.[3]

Ever since then work on exporting congestion by means of new streets has proceeded without interruption. Desires constantly under threat demand offensive defense: in the battle against traffic jams, bypasses and loops, expressways and highway routes sprung from the ground—stolen time would be made up for by increased bypass kilometers. And the old story is repeated with beautiful regularity: in no time at all, new strips of asphalt entice the bus and streetcar riders into cars, thus encouraging, over the long run, more car ownership and more moves to the suburbs. Nothing stimulated traffic like street construction. And as soon as the new loop suffered its first traffic jams the devaluation of kilometers began anew, right alongside the devaluation of minutes: since traffic moved more slowly on the newly built bypasses, drivers now took longer to cover the loop's longer route to the office. If mass motorization has the—inappropriate—result that more time is needed for the same route or less distance is covered in the same time, then everyone is worse off than before. No wonder we hear the beams creaking in the house of desires.

Solitude in a Crowd

Much as the outwitting of time often ends in congestion, the urge for the romantically remote often discovers only the old and familiar. The automobile made the longing to flee the confines of daily life in the city—off to untouched nature and unspoiled customs—into a general good. With mass motorization the saga of popular wanderings now got underway, the four-wheeled search for blue blossoms. The utopia of the leisure society, where the common man, though inextricably tied up in wage labor, also enjoys its opposite in the evenings and on weekends, by being able to move agilely from the office into the mountains, from the firm into the woods—this utopia is closely related to the automobile and drives many onto the roads. In 1975, approximately 80 percent of leisure and vacation travel was accomplished with the automobile.

Admittedly, mass motorization has changed the rules here as well: just as the advantage of being faster than the others has shrunk, so has

3. *Berliner Illustrirte Zeitung,* 1926, no. 21, 645.

the advantage of being able to travel farther. The attraction of a distant locale—whether a still pond or a lonely stream—is founded on the expectation that it will be accessible to only a few people. As soon as millions of weekend nature lovers start strapping their camping tables and surfboards to their car roofs and go swarming off on the search for solitude and the pleasures of the landscape, all of them involve themselves involuntarily in destroying, simply through their presence, the very solitude and pleasure they all seek. "It is not yet known," wrote Jürgen Dahl in *Die Zeit* in 1971,

> to what extent a people is really willing and able over the long run to forgo traveling in their automobiles to the last refuge of quiet, to the desired oasis, destroying that oasis with exhaust and noise. Flight, even when it is so well prepared, always remains a last resort. Organized mass flight can only end like the story of the hare and the tortoise: when all the refugees have reached their destination, they find there that which they wanted to flee.[4]

The attraction of the distant locale relies on exclusive accessibility; it deteriorates when hordes of automobiles invade it. Streets line the seashores and parking lots fill the mountain fields; the world that counters the industrialism of daily life is itself being industrialized. Vexation and disappointment are unavoidable. Nor is the automobile-centered suburban lifestyle immune: as soon as all the others move in, the wilderness slips farther away, and so looming urbanization pushes the next development even farther into the country surrounding the city. And of course the immediate solution—to retreat from the urbanized countryside by fleeing to ever more distant destinations—only leads to another dead end: space becomes completely bereft of secrets, and no more surprises lie in wait for the tourist's gaze.

Because the automobile loses its advantages as it becomes democratized, the social limits of advancing automobile consumption, like the physical limits of energy and clean air, pressed to the fore in the 1970s. Not only is the supply of gasoline, and of breathable air, being exhausted, but desires are becoming exhausted as well. With mass motorization individual aims that promise advantages quickly add up to a social sum with fatal consequences; thus the original aims are turned into their opposite. In a kind of counterproductive collaboration, although each individual acts rationally, all of them together act irrationally. This dilemma becomes evident in manifold experiences and eats away at the hopes that once surrounded the automobile. A series

4. *Die Zeit,* 1971, no. 28.

of Allensbach surveys found that in 1960, 63 percent of car owners derived great pleasure from driving; in 1981, however, the figure stood at only 41 percent.

Smith and Jones Catch Up

In 1911 the membership list of the recently organized new German automobile club read like a "Who's Who," teeming with bankers, manufacturers, and military officers. Car ownership indicated a person's position in society. Those who were graced with earthly goods displayed their automobiles to communicate with conspicuous inconspicuousness where they belonged and what worlds separated them from the man on the street. To imitate the wealthy and distinguish oneself from the eager have-nots—such was the motivation that gave the car a larger-than-life image and sank the roots of longing for an automobile deep into popular sentiment. In the 1950s particularly, the automobile was the focal point of many hopes for upward mobility, and the statistics prove that income groups, one after the other from top to bottom, outfitted themselves with cars throughout the postwar years. Does it not do irreparable damage to the attractiveness of the automobile if the few lucky ones turn into a whole nation of drivers? It was with some melancholy that a journalist in 1981 glanced backward:

> Certainly, it all began rather slowly, but that simply made it more tantalizing: automobiles were there to be seen, unattainable at the outset, being beyond the means of those whose first concern was simply to pay the grocer. A little later one could at least begin to calculate how long it would take to accumulate the necessary savings, and the dreams became a little more concrete. Finally, somehow, the time came: . . . the car sat in the drive. Now real life could get underway: wash the car, wax it, polish the chrome on Saturdays, and then a weekend outing; enjoy the envy of neighbors and work associates; comfort the children when they'd gotten a punch in the nose from their jealous, carless peers. The world of the automobile seemed holy, the sector became the leading economic indicator. . . . It just isn't that way anymore. One holy world after the other turned out to be a soap bubble. Even overhead cams can't make up for it in the end.[5]

With mass motorization, the one-time luxury item has sunk to the level of a universal commodity. The exhilaration is past and daily life resumes: that banal piece of standard equipment that is now part of every household no longer gives cause to turn up one's nose. Because

5. *Vorwärts-Spezial*, August 1981, 18f.

the automobile is not in short supply anymore, its simple possession no longer serves as a distinction. It has lost its differentiating power as a symbol: the desire for social superiority no longer finds in the automobile as such a means of gratification.

Yet, when everybody owns a car, the kind of car one drives can become important. The increasing differentiation since the mid-1960s of car types according to performance categories proves that the desire for social distinction has shifted its target: now the urge is to own a higher-class automobile. A big, high-performance car can at least secure little victories on the street; by allowing owners to mark the difference between themselves and the rabble, it displaces social superiority into spatial superiority. "At 140 kilometers per hour," an advertisement for the 1970 Citroën DS began, "the DS regains its distinction; at 140 kilometers per hour everything begins to return to its proper order. The fast lane is once again reserved for the DS, leaving others to look on as it quietly but inexorably pulls away."

Admittedly, the joy of high-performance cars did not remain untroubled either. In 1963, drivers of cars exceeding 1500 cc could still feel special, since only 18 percent of cars had such power; by 1979, however, this advantage had melted away, for 50.5 percent of drivers could boast of such a car—any question of exclusivity was moot. The symbolic value of a big car has also been undermined by inflation: those who moved up to a higher performance category to preserve a social distinction soon realized that they had been fooled, for the competitors—their social inferiors—were also seeking to reduce the gap. To maintain their relative position, then, they had to exert themselves all the more, to walk up the down-escalator, so to speak: not to move forward is to fall behind. Thus high-powered motorization lost the power to mark difference—especially since the underdogs, enjoying ever more purchasing power, could so easily manipulate the symbol themselves: then the socially disadvantaged end up driving the biggest cars.

Mass motorization prepared the way for the status-conscious, beginning with the upper strata, to turn from automobiles and focus on other prestige items, like airplane trips and second homes. The charismatic force of big cars declined; a smaller car does just as well these days, and an enormous automobile no longer demonstrates anything but status anxiety. Even advertisements allude to this unspoken consensus, like one showing a doctor and a pretty woman in front of the hospital talking about his rather modest Fiat Ritmo. She asks, "Why don't you drive a prestigious car like the other doctors?" He answers, somewhat condescendingly, "Presumably because I'm a psychiatrist." There is relief for the psyche, then, for those—even doctors—who can find their casual satisfaction in a Ritmo. The hope that for half a century now has been attaching all those dreams to the automobile is used up: the possession of a car lifts one above the Smiths and Joneses about as much as does the possession of a vacuum cleaner.

The aging of desire—is it the result of a dialectical process, in the true sense of the term? The cultural attraction of the automobile was so strong that it created the material conditions that now undermine the original attraction. Mass motorization depreciates the earlier hopes and looses contrary experiences that grew in protest against a society built for the car. With the breakdown of the old consensus during the 1970s, the automobile has become a bone of political contention. Caught in the trap of disenchantment, attitudes toward the automobile become polarized. One response is to grow desperate about achieving happiness: to attack the problem of urban congestion by building wider streets, the overcrowding of the landscape by opening up new regions, the loss of prestige by fleeing to ever higher performance engines. But another suggests that we abandon the old dreams: to discover, in the face of congestion, the qualities of prudence; to reconstruct, in the face of overrun distant attractions, our sense of home; and to glimpse a new privilege of status in the sovereign renunciation of the automobile.

The Dominion
of Long Distances
and Speedy Execution

In Greek mythology, an ominous figure ready to prove that pride goes before a fall is always present: those who tempt the gods entice Nemesis, the goddess of vengeance, into the affair, who causes their haughty behavior to rebound to their detriment. Icarus (included, contrary to the sense of the myth, in occasional triumphal histories of transportation as the first flight engineer) learned the horrible lesson of his limitations: the wax holding the feathers of his wings together was no match for the hot rays of the sun, and he crashed and sank into the sea. Looking back at the history of automotive enthusiasm, it seems the vengeful hand of Nemesis has been at play: what began as a grandiose advance toward liberation ended in a finely woven net of dependencies.

It was not that the longings for automotive freedom were wrong, but that the promises of motorization proved deceptive. The automotive compulsion grew alongside the liberation gained from the automobile: even if the desires have been used up, we cannot easily choose to do away with the automobile, because it has been transformed from a luxury item into a piece of equipment presumed necessary for survival. All the doors seem to have slammed shut; none offers a way out of the transit-intensive society. Does then—and this question is asked especially by the young—the future have a chance?

In Remembrance of Time Gained

Of drivers responding to the Allensbach survey of 1981, 36 percent took no great pleasure in driving; they were, they indicated, on the road because they had to be. The enthusiasm has gone flat; the dictate of necessity has pressed to the fore. For many, it seems, the automobile has changed from an article of pleasure to an enforced article of utility. Whatever the individual motives for owning an automobile, drivers, against their will, are nothing more than the inheritors of the hope of "covering ever greater distances in ever shorter amounts of time" by means of cars.

This prospect, it will be recalled, gave wings to the early "automobilists" and later enticed many to play a trick on fleeting time and obstinate distance by buying an automobile. But success here proved to be a boomerang: the jump on time and distance that car owners once enjoyed has melted away with mass motorization. Even worse: what was once a jump has now become so common that there is scarcely any escape from its domain. Everyone assumes that the other has access to a car and so silently expects auto-mobilization: the school transfers remedial instruction to the afternoons because the teacher has no trouble returning; the therapist sets up shop in a nice area away from the city center because clients all have cars anyway; and the plant demands punctual arrival at an inopportune time because the workers can get there early enough by car. Time has certainly been gained with the automobile, but the gain is no fun if it becomes an obligation. Today's obligations are made of what formerly were desires. What was once a head start has been generalized and incorporated into the rhythm of daily life. In a society whose time regimen is based on the automobile, it is no longer true that people take the car to work; rather, it is the latter that puts them to work.

The time gained, however, has not disappeared without a trace, though one certainly does not find it where one most expects. Contrary to popular belief—and this is proved by a multitude of studies from many countries—drivers do not spend less time in transit than nondrivers. Both groups expend an average of seventy to eighty minutes every day on their mobility. Nor are drivers more frequently on the move; they leave the house slightly less often than nondrivers. Those who buy an automobile, that exalted "time-saving machine" of the early years, do not take a deep breath and rejoice in extra hours of leisure; they just roll over greater distances when they do go out.

Rather than saving time, the automobile makes it possible to seek out more distant destinations; its powers of speed are cashed in not for less time on the road, but for longer routes. Having a car in the garage invites one to find a job in the next county instead of at home, or to prefer the club downtown to the local bar. The driver's gaze squints into the distance and selects a destination well over the horizon. The driver's lifestyle consumes space. Whether purchasing some furniture or simply going swimming, drivers, as if in seven-league boots, traverse an ever wider range. Motorization exploded the arena of activity. The time gain is reinvested in longer distances.

Mobility-Consuming Distances

All of this seems fully within the spirit of the hopes vested in the automobile, but in hindsight we can recognize the blind spot in the utopia of ever-increasing mobility. The places where we meet friends, visit the doctor, or simply buy a loaf of bread have not held still but to a large extent have wandered off themselves. Residential arrangements do not remain the same but grow along with the exploding radius of activity. The city, and the village too, are less concentrated, with the result that destinations once located in the neighborhood are moving out along an ever broader circumference. Whether we want to do our shopping, visit friends, or go to work, we have to bridge longer distances, for the mom-and-pop grocery has disappeared, the friends have moved to satellite cities, and the plant is now on the other side of town.

Life is being broken up into space, because individuals and institutions alike often base their location decisions solidly on the assumption of the automobile. In Hamburg, for instance, between 1959 and 1979, 226,000 people moved from the core city into the suburbs, purchasing proximity to nature at the cost of a longer commute. Things are combined, consolidated, centralized, and rationalized, and shopping centers, like magnet schools, leisure centers, and industrial parks, contribute to the thinning out of neighborhoods, forcing customers out onto lengthy journeys. The cities and rural areas are being reconstructed in such a way that soon everyone will need an automobile. Exploding distances allow for practically no other choice than to live a transit-intensive lifestyle. Everywhere people are traveling farther—between 1960 and 1980, the average distance covered each day rose from 13.2 to 22 kilometers—but whether according to desire and fancy or duty and obligation remains to be seen.

If among these extra kilometers some are made necessary because what was once nearby has now moved away, is it not nonsense to celebrate increased mobility as an increase in the quality of life?

> When the milkman no longer comes to the door, it is a dirty trick to portray the possibility of going oneself to fetch the milk as desirable mobility. And the freedom to choose any conceivable mobility in the exercise of a profession only means that, rather than moving on the labor market, one is instead moved by it, and in the most dreadful way. . . . The supermarket in an open field far from the city first became possible through private motorization; if it did not exist, we would be supplied by private grocers, who are now in agony on account of motorized competition. And so on. The possibility of transportation has become the necessity of transportation. The freedom to change places has produced the compulsion, if you will, to travel to pursue a profession, or to acquire goods, or to do anything else, all of which not long ago was accomplished with a minimum of movement. . . . That Immanuel Kant never left Königsberg in his entire life renders him pitiable to ridiculous in the eyes of our fast-paced contemporaries, because they fail to consider that to stay in Königsberg in the eighteenth century was altogether less tormenting than being chained in the twentieth century to Düsseldorf.[1]

The dream of gaining freedom through mobility thus ends in a vicious circle: one acquires an automobile—one travels farther—important destinations wander out of the vicinity—they can scarcely be reached without a car—and so others see themselves compelled to switch to the automobile . . .

A Nation of Passengers and Commuters

The lengthening of obligatory journeys has produced something approaching an anthropological mutation: for a generation now, many have found themselves scarcely able to make it through the day without being strapped to an engine. A new basic need has taken its place next to such venerable ones as clothing, food, and shelter: the comprehensive need for transportation. Such a pass was unthinkable even for our grandfathers, who, unless they lived in a large city, undertook at most an occasional trip by rail but otherwise managed the range of their daily world with the help of horses or their own two feet.

Whether one is rich or poor, whether an automobile driver or a bus rider, it makes no difference. If 77 percent of all people working out-

1. Jürgen Dahl, *Die Zeit*, 1973, no. 46.

side the home in 1982 got to work by means of motor conveyance, then practically everyone has become a prisoner of transportation. Given this fundamental shift, the rumpus between those who champion the private automobile and those favoring public transit seems like an argument over the superior path into the same imprisonment. The traditional leftist utopia of publicly organized transit (preferably free of charge) would indeed be more democratic, but it would result in a sort of democratic despotism: all would be equal in their dependence on a transportation system.

With its rapid development into a means of mass transportation—a decisive transformation—the automobile has lost its one-of-a-kind dignity and become a mere cog in a comprehensive machinery. As a result, its aura as a means to gratification has been lost as well. The final judgment is sober: the private automobile represents merely a mode of access to a social transportation machinery whose logic compels journeys having nothing in common with the pleasure trips of earlier years. Long distances and the accelerated pace of daily life often leave one no choice but to acquire a car. In a society centered around the automobile, the purchase of a car easily becomes a matter of self-defense, a way to avoid—whether on the labor market or in private life—a loss of status. More cars truly do not mean a higher quality of life. How many new registrations in fact signify acts of self-defense, aimed at maintaining one's position, rather than stabs at gratification, aimed at getting one step closer to the good life?

The promises have become obligations, and the one-time instrument of gratification a means of self-defense. The masters of space and time awaken to find themselves slaves of distance and haste.

The Decline of
Autonomous Mobility

The more zealously automobile enthusiasts recast the world according to their view through the windshield, the more emphatically is forced to the fore an insight that is actually nothing more than a platitude: in short, even the most passionate driver will sometimes go by foot.

However uplifting sitting behind the wheel might be, once there are twenty-four million automobiles on the road the train of consequences that each vehicle drags with it must be seen vastly to exceed the dimensions of the car itself. Unlike a vacuum cleaner or an electric razor, the automobile is an apparatus whose operation implicates nonpartici-pants; it produces, as the economists put it, obligatory exogenous effects. Noise, pollutants, accidents, the consumption of space—by now every child can repeat the automobile's litany of sins: no more green on the trees, and frogs end up flat as flounders; lead in the blood and noise rattling the nerves. The dreams of yesterday have produced the nightmares of today.

The polluting of the social environment will eventually catch up with the polluting of the physical environment. Streets disrupt our walks, and intersections annoy our eyes; composure is undermined by the hectic, and the homeland unravels into distances. Making cars also leads to the breakdown of environmental conditions in which a non-motorized lifestyle flourishes. The compulsion to be on the move has so changed city and countryside that simply being there is no longer any fun. Why else is everybody always on the road?

The Uninviting Here

The domination of distance has brought with it a revaluation of space: the gaze is directed at distant locales beyond the neighborhood, and the immediate spatial world has declined in significance. Yet the attractiveness of the distant is purchased at the cost of a demotion of the near. That which is gained at a distance is often lost close to home: in villages and city neighborhoods the opportunities to shop, to have repairs done, to meet friends, or simply to let oneself be carried along in the movement of the world have dwindled. Instead it has become harder for elderly persons to visit their doctors, children no longer have fields and streams in which to play, and housewives find themselves left behind in bedroom communities.

The "emptying" of the villages and the "desolation" of the inner city have become virtual clichés, formed in the slipstream of the automobile's greed for distance. The part of society that is visible with one's own eyes, accessible on one's own feet, is smaller. Nothing happens in the neighborhood anymore: the corner store has capitulated to the shopping center on the outskirts of town; the local dressmaker has lost her customers to the department store downtown; people desiring a walk hurry off in their cars to the nature paths; and a driving school moves into the corner bar. The local area, once easily mastered on foot or with a bicycle, thins out and has less and less of interest to offer. The social fabric held together by foot traffic tears apart. Thus we note a paradox: the motorization of society, the project of quick accessibility for one and all, has pushed an important piece of the world away and made it—especially for those who have no automobile—less accessible than before. In parallel with the mobility of the motorized rose the immobility of the nonmotorized.

Yet there is more: the immediate vicinity, with its nooks and alleys, its stubborn particularity, is blocking the way. It puts a brake on the speed of those who, exulting in motorized power, want to rush along the fastest route to their distant goals. Precisely where the neighborhoods are tailored to the short and winding paths of pedestrians (which was, indeed, the point of the medieval city structure, with its tight network of narrow ways), that is where things are excavated, shoved aside, evened, and unified, until the village green is made into a thoroughfare, and the beer garden becomes a parking lot. Penetration in the straightest line possible is what counts. War has been declared on the niches and yards, the narrow streets and little squares, on those places where the subtle texture of neighborly relations makes its architectural stand. For the gaze greedy for distance, the living space of the immediate vicinity degenerates into mere thoroughfares, into a dead space between the beginning and the end; the point is to overcome this space with the least possible loss in time.

The results of this rage for passing through are everywhere visible. On streets that, not so long ago, were graced with trees and lined with little shops and cafés, populated with playing children and chatting neighbors, dozing grandparents and busy passersby, now there is tin, concrete, and endless tumult. The space available for children in particular has been reduced so drastically that today the order "Go to your room!" has largely replaced the earlier "Go outside!"

Urban living spaces have become transit strips with waiting areas. Built in the name of forging quick connections between people, through roads have led instead to separation. Because each new thoroughfare cuts not only through residential areas, but also through contact networks, it can become an unsurpassable hurdle, especially for children and the elderly. In any case, main arteries accelerate through traffic at the expense of pedestrians and bicyclists, who have to look for overpasses and wait for traffic lights to change. In addition, of course, the inhabitants are left with noise, filth, and the increased risk of accidents, not to mention their justified complaints over the modern form of land theft: in 1974, 14.5 percent of the total area of Frankfurt was given over to traffic, as opposed to 6.9 percent in 1950. Thus is the immediate vicinity placed in the service of distance and degraded into a space for transit.

Who owns the roads? When this question was decided in favor of the automobile in the first decades of the century, no one could have known that one day the car's dominance would be poured over the streets in concrete. As long as only public ordinances maintained the primacy of the automobile, as long, in fact, as the naked power of the engine paved its own way, this dominance was controllable. But today, now that it has been institutionalized in expressways and residential structures, that dominance seems almost natural—indeed, some streets are no longer fit to be anything but a space devoted to transit. Not only are the nonmotorized not allowed to appropriate the street space, but—more portentous yet—they are no longer *able* to. Where space is carved up according to the automobile's demand for passage, little is left for the pedestrian to experience, see, or do. Space conformed to speed destroys space conformed to the pedestrian.

Pedestrians (and bicyclists) love the minor and incidental. They feel good where the buildings wear different faces, where the eye can wander over trees, yards, and balconies, where there are people to meet or watch, where they can linger, join in, and get involved, where a multitude of impressions and stimuli can be had along their short way. The area one can cover by foot corresponds to space that is closely knit, multifarious, and rich in events.

The situation is wholly different for drivers: they hate surprises and demand predictability; only drawn-out monotony gives them security; only large billboards can capture their attention; only straight, broad, and uneventful routes guarantee them a quick passage without interruption. The car driver tolerates variety only in the rhythm of kilometers, whereas for the pedestrian, space made to conform to speed is faceless and boring.

The automobile contributed significantly, of course, to the ruination of the structural and social "ecosystem" in which pedestrians and bicyclists feel at home. The pedestrian needs a thick, intertwined, even entangled locale. It is not without reason that places built by their inhabitants to their own measure often resemble labyrinths—one thinks here of a Moslem medina or a medieval city. The labyrinth is the ideal structure for a people who rely only on the power of their own legs: it encompasses in the narrowest possible space a multifaceted world and creates security for those who spend their daily lives there (if confusion for strangers).

The opposite of the labyrinth is a space planned for the automobile; with a priority on rapid through transit, no environment hospitable to the pedestrian is possible. The most decisive consequence of motorization is the destruction of the vital basis for nonmotorized movement— and this goes for the all-around "clean" car as well. As a saying current in Los Angeles goes, "Pedestrians are people on their way to or from their cars." The automobile has arranged for itself a radical monopoly, one that causes not other firms, but entire ways of life to disappear. "This profound control of the transportation industry over natural mobility," Ivan Illich wrote in 1974,

> constitutes a monopoly much more pervasive than either the commercial monopoly Ford might win over the automobile market, or the political monopoly car manufacturers might wield against the development of trains and buses. Because of its hidden, entrenched and structuring nature, I call this a *radical monopoly*.[1]

Those in the Shadows . . .

The right of unfettered movement has, in the wake of motorization, been transformed into an obligation toward transportation. Indeed,

1. Ivan Illich, *Energy and Equity* (New York: Harper & Row, 1974), 45.

transit-intensive distances and the uninviting here have created an environment in which the nonmotorized can scarcely survive. But what happens to those people?

To be sure, individuals without a car are not well off: they have the choice of either taking the time and trouble to use mass transit or not going anywhere at all. Many a grandpa in the countryside will just shrug his shoulders in resignation when he wants to buy shoes or visit the doctor. Lacking a car, he has only the unpleasant alternatives of rocking along in a bus for half a day or just staying home. Not much is available within the range of his own legs, and important destinations have been pushed too far from home. Those without an automobile find their power over the space for which no car is needed devalued, while their access to the space outside this narrow range is withheld. Motorization has created a new form of inequality.

In a pedestrian city in earlier years, say Tübingen in the nineteenth century, just about everyone but the lame had the same power over space, because all—with the exception of coach owners, and even there the discrepancy was not so great—were subject to the standard set by their legs. With motorization, the dominant classes acquired another means of exercising power over space, and accessibility—formerly generally available—became a scarce good that could be had only through the purchase of transit kilometers. Thus was laid the foundation for transport-based technological inequality: the better off grabbed the newest means of transportation more quickly than the less prosperous could keep up. The gap between the privileged and the unprivileged widened—to the increasing disadvantage of the nonmotorized because, in the struggle between the two groups, autonomous mobility was the inevitable victim.

For those who insisted on relying on their legs, the imperialism of the motor soon demonstrated their handicap. A reporter detailed her bicycle trip through Frankfurt under the title "Lost in an Avalanche of Tin":

> The first dilemma [comes] on Berliner Straße at Kornmarkt. Bicycle riders always ride on the right, but the right lane goes around the corner. So it's straight ahead into the speeding traffic of the other two lanes. Ride around the Theatertunnel—a blue sign says it's reserved for cars. The only thing that prevents me being run down by the traffic turning right is braking to a halt. A gravel truck goes by, little rocks flying. . . . I can't disappear into thin air, even if the plaintive drivers

seem to be demanding it. So I somehow squeeze into the middle lane, mentally pulling in my head.[2]

In fact, the 1913 call to protest by the Freiherr von Pidoll, who warned that general use of the streets would be destroyed by the monopolistic ambitions of the automobile, had a nearly clairvoyant quality: those who do not succumb to being driven off the streets risk life and limb. In 1980, 56.3 percent of local traffic casualties were pedestrians or bicycle riders; whereas on average only 3 percent of pedestrians came away without injury, 79.5 percent of automobile occupants were unharmed. Reliance on leg power, moreover, is much more in vogue than the view from behind the windshield would like to acknowledge: in West Germany in 1976, 40 percent of all movement from place to place was accomplished on foot or by bicycle (not counting children under ten years of age; should we include them, approximately half of all such movement—including long-distance transit—would have involved no use of motorization).

Why, despite all the dangers and repression, is nonmotorized traffic much more substantial than motorization, in its rapture, suspects? The answer is as simple as it is surprising: only a solid third of West Germans have constant access to a car. In 1978, 18.6 million private automobiles were registered, in a population of 62 million. Although in 1979, 61.8 percent of all households had a car in the garage, many households consisting of a single person or elderly individuals or earning little income often have no horsepower to which to resort. Even in motorized households, not every family member manages to get behind the wheel; children and teenagers are excluded first of all, and not every adult has a driver's license. Moreover, men set the tone with automobiles: only 17 percent of all men had no driver's license, as opposed to 58 percent of women. By including the undermotorized groups in our calculations, we arrived at a simple fact: the majority of the population lives at the dead end of motorization. Those who reap the benefits of acceleration are predominantly employed twenty-five- to sixty-year-old males. Children, teenagers, the elderly, housewives, and people with low incomes profit only marginally.

While the motorization wave did substantially overcome class inequalities in car ownership (workers no longer fall dramatically behind civil servants, professionals, or farmers), it nevertheless reinforced a new type of inequality: the one between members of society regarded

2. *Frankfurter Rundschau*, 1980, no. 145.

> Walking. There are books in which words of similar meaning are grouped together. In one of these reference books the following appears under "Walking": "1. to put one foot in front of the other, move oneself forward, betake oneself, wend, proceed, take a road, cover, set foot on, or traverse; 2. to step, wander, travel, stroll, go for a stroll, jog, trot, roam, trip, trip along, amble, stride, stalk, tippy-toe, glide, tap, feel one's way, toddle, jog along, dawdle, trample, step, stamp, patter, shuffle, pad, ramble, slouch along."
>
> D. Garbrecht, *Gehen. Plädoyer für das Leben in der Stadt* (Walking: An appeal on behalf of city life) (Weinheim, Basel: Beltz, 1981), 186.

as productive versus unproductive. Motorized power over space is distributed according to one's proximity to, or distance from, the money-generating production process. Those who do not work in the service of output, well, let them move more slowly. Some are accelerated, others slowed down; time saved for one person goes on the books as time lost for another. The net result might be the same, but the opportunities for mobility have become polarized—and in no sense have they increased across the board. In the myth of the "automotive society," the minority of financially productive people are taken to represent the whole of society. No one sees those standing in the shadows.

Tranquillity Rediscovered;
or, For Love of the Bicycle

Images of the future acquire their coloration in contrast to the present. The automobile captured hearts initially because it carried the light-footed promise of overcoming that which weighed contemporaries down: the fear of being chained to one spot, of being bound to daily life and at the mercy of frail corporeal power. Automotive dreams lived on the contrast with an immobile world, on the fear of being tied down; only in the soil of a social claustrophobia could the rage for mobility flourish.

Such has not been the case since the 1970s. It is no longer fear of immobility that nourishes alternative designs for the future, but the anxiety of homelessness. In a society on wheels, the promise of more speed or more engine power easily loses its attraction; where vehicles are everywhere but no one arrives, where everyone is underway but no one connects, other desires take form. In his book *Ecotopia,* Ernest Callenbach sketched a postautomotive utopia:

> I checked my bag and set out to explore a bit. The first shock hit me at the moment I stepped onto the street. There was a strange hush over everything. I expected to encounter something at least a little like the exciting bustle of our cities—cars honking, taxis swooping, clots of people pushing about in the hurry of urban life. What I found, when I had gotten over my surprise at the quiet, was that Market Street, once a mighty boulevard striking through the city down to the waterfront,

has become a mall planted with thousands of trees. The "street" itself, on which electric taxis, minibuses, and delivery carts purr along, has shrunk to a two-lane affair. The remaining space, which is huge, is occupied by bicycle lanes, fountains, sculptures, kiosks, and absurd little gardens surrounded by benches. Over it all hangs the almost sinister quiet, punctuated by the whirr of bicycles and cries of children. There is even the occasional song of a bird, unbelievable as that may seem on a capital city's crowded main street.[1]

People—minorities to be sure, but eloquent ones—outraged over leveled forests and pierced-through neighborhoods, have fought during the last two decades with reports and demonstrations against the paving of the landscape ("The State of Hessen, in Contract to the Federal Republic of Germany, Is Paving the Remainder of Same"). In doing so they have lent legitimacy to new social images that have little to do with the automobile. This historical transformation found its voice in popular referenda.

Only a Shrug for Progress

That the promises of yore have lost their value, that mass motorization conjured up a web of compulsions and transportation obligations, that the world became less practical for people who did not sit behind the wheel—these experiences have been accumulating since the 1960s and cannot help but pull the rug out from under traditionally triumphal automobilism. Admittedly, not everyone has antennae for the transformation. In the referenda, the well-educated youth were essentially unanimous, with the twenty- to thirty-five-year-olds with a high school or college education largely forsaking the faith of their fathers. The 1960s were, after all, their politically formative years—hence their reaction not to the shortage, but to the excess, of automobiles. Those children of the economic miracle with the most hope, in particular, had nothing but scorn for their fathers' jubilation over growth. Similar historical experiences marked their views, making them more than a mere age group, but a social group, united in a mutual, revisionist sense of life: "Progress? No thanks!"

Arousing the disdain were precisely the persisting signs of the former enthusiasm: a new stretch of highway inspired acrimony, and the occasional felling of trees for a road drove even dutiful citizens into the arms of the environmentalists. The automobile advanced to become

1. Ernest Callenbach, *Ecotopia: The Notebooks and Reports of William Weston* (Berkeley: Banyan Tree Books, 1975), 11.

"Environmental Enemy Number One." When such a contrivance blows a whole kilo of lead into the air for every hundred thousand kilometers driven; when the thoroughfares around Lake Constance turn the landscape into islands 6.3 square kilometers large on average, such that a pedestrian confronts another interchange after a stroll of only three kilometers, at most; when in West Germany nearly half a million people have been sacrificed on the altar of the automobile—then the faith in progress that gave wings to the cultural rise of the automobile must one day collapse. In the process, conflicts over the preservation of this urban greenbelt or that half-timbered cabin pick up a dynamic far greater than the individual case demands; indeed, those taking a stand in front of the endangered trees are often people who before could not distinguish a beech from an oak.

Evident in these confrontations was a disenchantment with the belief in progress, held from the end of the nineteenth century up to our day, whereby the future, so full of promise, would always surpass the present, because all it took was a steady rise in the number of goods to keep history moving along its upward course. The felled tree and demolished cabin became symbols of the insight that this progress in fact relies on a disproportionate transfer of costs, in which, for the pleasure of the driver, society becomes a garbage dump for the by-products of motorization.

As the garbage rose, the quality of life fell. The conviction that "bigger, farther, faster" brought one closer to the good life, the obsession with the future—without which the rage for speed, not to mention the addiction to novelty and the stress, could never have blossomed—was over. On the contrary, it appeared that retrogression was marching hand in hand with progress, and technology seemed to liberate by subjugating. Still more fateful, the escape routes appeared to have been closed, the way back blocked, alternatives destroyed, the future colonized. Where the dictates of objectivity reigned, the future had played itself out, and with it hope as well. With society apparently condemned to total motorization, the only remaining choice was to manage it. *No future.* What the minorities were articulating seemed to be gaining ground even within the majority of the population: in 1956, according to an Allensbach survey, 56 percent still believed "that people are moving toward a better future"; in 1980, only 28 percent persisted in that belief.

But how greatly Callenbach's 1975 "Ecotopia" differs from the utopia of *Das Neue Universum* of 1913 (see above, p. 128–29)!

Whereas back then one was confronted with images of a world riddled with tunnels and enveloped by air routes, and fantasy was dominated by a fully penetrable space, today more contemplative images are common: apartment blocks surrounded by green grass, bicycles, fountains, and sculptures—and, naturally, the singing of birds. In a crisis, creativity blossoms with notions anchored in the romantic tradition. The desire for a life maintained intact from the overweening power of appliances—that is the leitmotif running through the traffic referendum movement, as well as the antinuclear and peace movements. The first item of debate is no longer how the fruits of progress are to be distributed, but how the colonization of nature and human life can be checked. "Nature," "health," "autonomy" have become keywords in the struggle against the devouring dictates of objectivity. Whether for the protector of forests, the health food enthusiast, or the grass-roots volunteer, maintaining a bit of life untainted by the economic imperialism they find at home is what matters. The history of desires takes leave of the history of automobiles. As for those who still cannot tear themselves from the wheel, their sentence is to suffer from a bad conscience.

Motorless Autonomy

The new aspirations for a gentler society have nothing in common with the automobile; they are much more attached to the bicycle. Are cyclists not masters of their mobility, without at the same time spreading harm to others? The bicycle, once the Cinderella of transport, suddenly rose to prominence as a symbol of humane technology; with its chain, ball bearings, and light metal frame, it offers the gains of advanced technology but does not threaten the environment with poison, damage, and dispossession. Moreover, it stands not only for undamaged nature and intact human beings, but also for unbroken autonomy. To attack the pedals may be strenuous over the short run, but it is also an expression of trust in one's own powers, for with the bicycle everything depends on the self: body and brain are not installed inside a comfortable housing and switched off. The political idea of bicycle riding calls on nature and physicality in order to gain a substantial moral basis over the claims of the machine. Those people, so goes the cultural design, who wish to control the direction of an aspect of their own lives and move beyond existence as mere clients and consumers— those people ride a bike.

"I was fed up," explained Carol Carl-Sime of Hamburg in 1982 in the magazine *Brigitte,*

> driving fifteen minutes in the car every morning to the office and then just as long around the block searching for a parking place. I said to myself: Hamburg is flat and has relatively many bike paths. With a bike you can take off when you want, don't have to sit around in nerve-wracking traffic jams. You arrive at the office refreshed from the movement and save time and money to boot.[2]

Suddenly, with the bicycle, age-old motivations return: taking off when you want; making little detours and stops along the way; not having to sit around with the masses in irritating traffic jams—in the age of commuters and passengers, the idea of independence looks to the bicycle. No more clogged streets, no more being late, no more visits to the repair shop, and no more insurance premiums to pay: after mass motorization had stripped the automobile of its dignity as a unique item and degraded it to merely a self-propelled component of a transportation machine, experiential motifs that once defined the contrast between the automobile and the railway began to appear in the contrast between bicycle and automobile. Independent of traffic jams, highway onramps, service stations, and infrastructural planning, of pipelines and tankers—that is how the bicycle presents itself.

The automobile, it turned out, was only apparently auto-mobile. We might become ever more independent of one another with private automobiles, but only by becoming more dependent on the whole: an assassin needs only hit his mark somewhere in the Middle East for the supply of fuel to dry up. A quiet fear in the face of the power of this superstructure—a power as inexorable as it is safe from influence—lends the bicycle its aura of autonomy: this draft animal needs nothing, costs little, and is quick. Whereas the purchase of an automobile is a gesture of submission to the transportation machine, the purchase of a bicycle becomes a demonstration of trust in one's own powers.

Indeed, the renunciation of the automobile can hold its own even as a symbol of social superiority, as the fact that bicycles often serve as the background to product advertisements suggests. In one ad for Gauloises cigarettes, for example, a young man is seen astride his bicycle in heavy traffic, calmly smoking a cigarette. The text profiles a lifestyle without a car: "The man who travels slowly because he gets where he's going faster, who has enough personality to do without

2. *Brigitte,* 1982, no. 11, 85ff.

"The man who travels slowly, because he gets where he's going faster. Who has enough personality to do without horsepower. Who saves energy and gains strength. The man who loves the authentic—he smokes Gauloises. The unmistakable aromatic tobacco." Beyond the mobility compulsion: superiority through riding a bike.

horsepower, who saves energy and gains strength"—he, one infers, shows that he's different. In the face of congestion, the stress of time, and the general existence as a passenger, riding a bicycle acquires a whole new distinguishing power. Those who can afford the independence, the calmness, the physical self-confidence—in short, the nonconformity—of riding a bicycle manifest in doing so their distance from the masses of automobile drivers who cannot give up their cars. For the children of the mass-motorized society, who suddenly realize that they live under the dominion of long distances and speed, an air of sovereignty surrounds the bicycle. Those who ride bikes are truly their own masters. They can scoff at the transportation compulsion and remain unaffected by increases in the price of fuel. Moreover, if flexibility allows, they need not plunge into wage labor to save the money for a car; with a thirty-hour week, they can enjoy more leisure time as well! Now it is the lifestyle without a car that receives the garlands of scarcity. Where the majority are forced to drive automobiles, the exclusivity of those who do not wins a new power of attraction.

Recapturing the Neighborhood

In the 1920s Werner Bergengruen described the richness of perception on a bicycle:

> It vexes me that people pity me because I have no automobile, but only a bicycle. In fact, I don't want an automobile. . . . Speed means nothing to me, but the intensity of a trip everything. And to that belongs the experience of the tiniest differences, the nuances—the driver of an automobile grasps only the big transitions, or perhaps the crude differences. I experience every detail of the gradual transition from one kind of landscape or person to another.[3]

Those who swing these days onto the seat of a bicycle, senses deadened by the perspective through the windshield, are surprised at how multifaceted, how varied the world through which they ride appears. The bicycle rider discovers out-of-the-way paths and unexpected sights; space becomes accessible and perceptible in the wealth of its minor details. Behind the steering wheel one sees nothing, hears nothing, smells nothing; the perspective through the windshield kills space, makes it into a mere transit route. For the bicyclist, however, the nearby details

3. Werner Bergengruen, "Eine Lanze für das Fahrrad" (A lance for the bicycle), in *Lob des Fahrrads* (Praise of the bicycle), ed. Peter Francke (Zürich: Sanssouci, 1974), 28.

gain a sharper focus. It is not the gaze into the distance, but the attention to the immediate vicinity, that keeps pace with the bicycle.

The bicycle trip, moreover, has nothing to do with the straight, distance-covering lines of the automobile, but proceeds with little excursions and interruptions. The bicycle is more maneuverable and certainly faster than any other conveyance for door-to-door routes of four or five kilometers. "My bicycle ultimately gave me more freedom," a respondent in the *Brigitte* report said. "I say to myself more often: now I'll pick that up, because I'm going by there anyway, and while I'm in the area I'll visit my girlfriend too. With a car I would never do that, because I'd have to go around the block three times looking for a parking place."

Because the bicycle invites one to take possession of the world near-at-hand, it stands for a postautomotive ideal: the transformation of the immediate vicinity into home. Habitable streets, shops and businesses around the corner, green front lawns, roof gardens, and solar energy collectors—hopes are directed toward the ecological transformation of the city into a self-conscious space for living. They evidence a new preference for a centered life, the desire for a place of one's own, all images contrary to the frazzled lifestyle compelled by the automobile. The bicycle becomes the showpiece of a technology that requires local involvement. It calls forth the image of a localized democracy in which all—both rich and poor, young and old—can increase their mobility many times over that of a pedestrian without, however, limiting anyone else in their own freedom of movement. The bicycle renaissance bears witness to the search for a society liberated from an obsession with progress.

Prospects

No One Wants the Blame

No one willed the outcome, but everyone contributed to its cause. Under the slogan "Freedom of Choice in Means of Transportation," half the population pursued the promises of the automobile and switched. Everyone can give rational reasons for doing so, from Otto Julius Bierbaum in 1902 to an anonymous driver's ed student of today. But no one wanted to contribute to lung cancer, denude the pine trees of their needles, or keep the children locked indoors. Only the advantages count in considerations of private utility; they, after all, are subject to individual calculation, whereas the damages are distributed so thinly that they appear only on other people's accounts. The utilities are individualized, the damages socialized.

Although rational in personal terms, "free choice" proves nearly irrational socially. There is no guarantee—and this is the chronic error in talk of "free consumer choice"—that it leads to the happiness of the greatest number, when each person is seeking to maximize his or her own satisfaction. Even when we find ourselves up to our necks in harmful consequences, a kind of structural irresponsibility blocks the necessary change of course. The only thing those who renounce driving are guaranteed is that they lose an advantage; it is in no way certain that others will follow their example and so cause the damage in fact to decline. Indeed, probably the opposite is the case: if the number of cars on the road drops, then driving becomes worthwhile for others, and those who have forgone driving have accomplished nothing more than their own loss of social position! That is why so many, in spite of all the disenchantment, remain steadfastly true to the automobile. Individual renunciation carries with it no promise of social effects; at most the automobile opponent clears the way for the automobile enthusiast. Not even the health freak, who scorns pork because it is loaded with heavy metals, shies from taking a short car trip—thereby insuring that foods keep their lead reserves up.

Through the decades, transportation policy has accepted the challenge of accommodating the demand for private automobiles. Until the 1970s no attempt was even made to consider the social rationality of motorization and address the common good through a corresponding policy. Thus, the automobilization project ran aground on its own consequences. Every additional kilometer produces less and less benefit; it compels more people into transit, slows down the nonmotorized, and lets the mountain of garbage pile up. Every additional automobile takes away on one side what it promises on the other; every bit of growth destroys more value than it creates. A rise in the number of kilometers driven per person and per motor vehicle lowers the quality of life another notch—and that condemns more than just the transportation system to the state of a colossal running in place. At the same time, the way back is blocked, because driving has in many ways become a matter of survival. The desires of yesterday rain down on us as the compulsions of today.

The project that began nearly a century ago as a story of liberation is today wedged in between a transportation compulsion on the one side and mounting burdens on the other: it has become the story of dependency at an ever higher price. Waste disposal and remedial costs—from catalytic converters to noise reduction walls to reconstructive surgery—are on a steady rise. It will require considerable expense just to stop any further destruction of the natural and social environment. This dilemma between falling utility and increasing burdens, the dead end into which the promises of automobilization have led, forms the background to the conflict over the future of transportation and society. What should happen? The crisis is giving birth to new designs, which in the debate about the future will focus on far more than just the automobile.

Streamlining Through
High Tech

A more aerodynamic form and a light body made of high-tensile steel, aluminum, and plastics: that's how an automobile might look in the year 2000. It should offer more security, use less gas, and harm the environment scarcely at all. Maybe it will even have a square steering wheel fitted to the monitor in the middle of the steering column. This video screen will transmit to the driver all the data necessary for enhanced control. All the required switches will operate automatically. The vehicle will be powered by a ceramic or partly ceramic injection engine. This was the prognosis of Albert Kuhlmann, chairman of the TÜV in Rhineland, in a speech delivered to industry representatives in Aachen.[1] Experts even envision the automobile of the future with a built-in street atlas or road map, in the form of interchangeable cassettes, that will pilot drivers through the welter of streets. Such cars would roll down "automatic streets," which would guide traffic by means of microelectronic equipment.[2]

The automobile of the year 2000 will be fully equipped with technology; researchers and engineers bend over the culprit, trying to determine how it could be made harmless. The automotive industry itself comes to the rescue of the old desires. It invests intelligence and finan-

1. The Technische Überwachungsverein, or Technical Control Board, is the agency responsible for setting and enforcing operating standards for cars, electrical appliances, and so on.—TRANS.
2. *Frankfurter Rundschau*, February 18, 1984.

cial resources in fending off the crisis, and thereby makes capital—though that crisis was responsible for bringing them, as participants in its creation, to money and power in the first place. Once transit-intensive living conditions are regarded as unchangeable; once, too, the transportation-industrial complex—exporter, creator of jobs, engine of prosperity—is judged untouchable for the sake of the national economy, then only one way out of the growth dilemma remains: better automobiles and better traffic flow.

The Rational Automobile

In an advertisement for Siemens, the German electronics giant, cars dash past in every direction. A young mother is in the middle of the street, cautiously pushing a stroller in front of her. The child is roughly at the level of the exhaust pipes. The text: "People are still in danger from exhaust. But modern technology will help us reduce these harmful exhaust emissions. An electronic ignition system decisively lowers not only the level of dangerous pollutants in the exhaust, but also gasoline consumption. Electronics that improves our daily lives." The message is clear: clean air can be had through top-notch technology. And posing as the true defender of the environment, none other than Siemens.

While ecologists still decry the environmental crisis, industry has already begun to harvest honey from the destruction. A few years ago it seemed that the end of industrialism had arrived; but now the crisis appears to be spawning a whole new generation of technologies and bringing forth a new social project: the rationalization of industrialization through superindustrialization.

A change of theme in how car manufacturers represented themselves was observable already in the 1970s. Vehicles no longer basked in the light of power, speed, and prestige, but suddenly were vying over status as its own best savior. The ads betray the extent to which the customer's attention had shifted: the automobile companies now came across as the masters of the crisis, bringing the harmful consequences of automotive progress under control with a new round of technological innovations. In accord with the tenor of public discussion, vehicles were praised as low in emissions, secure, quiet, and, finally, fuel efficient; fuel injection systems and crash buffer zones, whisper-quiet engines and aerodynamics are the additions to the traditional technology that will attach the new expectations of the public to automotive product offerings. Performance at any cost was no longer the demand, but performance through reduced consumption. Body design mirrored the shift in aspirations as well: the wedge shape—slanting down to an edge in the front, with a high, squared-off back—became established as the basic form, standing for economical efficiency; in it, the ideal of the "rational automobile" gained palpable form.

With the rise of microelectronics as the technological basis for the anticipated superindustrial breakthrough, the chances of investing the automobile with the aura of economical efficiency also increase. The German electronics periodical *Elektor* muses over the historical mission of the microprocessor:

> Decades of technological progress according to the maxim "bigger, faster, costlier" has blessed us with a vehicle that, in the opinion of some, fits as well into the changed environment of the final years of the twentieth century as a dinosaur. Since we scarcely can (or want to) conform to a life without cars, however, the automobile must be made to conform to the changed conditions of the next decades. The automobile must be economical, safe, and clean, following the new demand. And as it happens, electronics has come far enough that it can be incorporated with great style in the automobile—which must be done if we are to save our beloved "dinosaur" from extinction.[3]

More horsepower and snobbish styling are yesterday's demands. The aging of desires has left its mark. On the agenda is not the expansion of force, but its clever utilization: the future lies in the inconspicuous elements beneath the tin. Computers take control where they can, measuring, calculating, adjusting. The chip portends a new age of efficiency; it promises an escape from poison and waste, the ills of a period of suffocating growth, by defeating them bit by bit from within.

Microprocessors make their way first of all into the engine and drive train, subjecting the machinery to demands for savings and cleanliness. The electronic injection system regulates the mix of fuel and air, and the electronic ignition the timing, such that no fuel is burned unnecessarily and fewer pollutants are released into the air. The two elements come together in the digitalized engine electronics, which is controlled by a microprocessor that continually monitors data such as oxygen used, temperature, and engine revolutions and computes control commands for fuel injection and ignition. The same is true for the

3. *Elektor,* April 1980, 24.

drive train: there electronics means that driving an automatic can be economical, because the computer shifts gears more intelligently and reliably than a human.

Microelectronics, in short, is busy removing the sting from the unpleasant experiences that led to disillusionment with the automobile. If no limits to growth had emerged, microelectronics would not have enjoyed such a status; as it is, however, this technology has gained a historical mission: to accost pollution and waste. Its message is decontamination and efficiency—but not only that: microprocessors also take responsibility for safety. Antilock braking systems prevent accidents by making lightning-quick variations in brake pressure to prevent wheel lock, and on their way are such things as a radar system to measure the distance from obstacles ahead—to meet the apparently urgent need to be able to drive at top speed through the fog.

Economical, clean, and safe—the message of modern automotive technology plays on the full scale of aspirations that arose in opposition to full motorization. Despite the surly polemic against environmentalists, automobile manufacturers exploit the historical break in the popular experience of the car: by redefining the traditional automobile as a source of poison and danger, they can sell their newest technologies to the fanfare of environmental protection. Taking responsibility for our environment does not demand a reduction in the number of cars; rather, as one advertisement claims,

> taking responsibility for our environment demands the newest technology. BMW delivers it. . . . To decide on a BMW has always been a sign of critical expertise. Today it also proof of an alert environmental consciousness. Exclusivity in automobiles means much more today: the critical consciousness must take its place next to high expectations. It has therefore never meant more than it does today to step up from standard solutions to a more intelligent technology, that is, to a technology stamped by an awareness of responsibility. . . . Nothing is more exclusive today than the maximum in efficiency.

This model of the BMW driver would be inconceivable without the sobering experience of the 1970s. The boundless expansion of desires is past; what counts now is their consolidation in a reduction of harmful effects. The disenchantment of triumphal desires in no way plunges the automotive industry into the void. On the contrary, the ecological skepticism of the seventies serves as a means of cultural transition to the conquest of society by a new generation of innovations. Microelectronics becomes the technological answer to the many-sided crisis of growth. The sign of superindustrialization is not the unconcerned plundering of natural resources of old, but has at its core the project of getting as much as possible out of limited resources. It aims at draining the flood of environmental burdens by means of countless canals, hoping in the process to rescue desire.

The Intelligent Highway

The computerization of society, ultimately a technological-cultural salvage project, is presented as the sanitizer of the old industrialization. Industrial growth depends essentially on attaching more and more needs to the output of goods and services supplied by factories and authorities—electricity instead of coal, potato flakes instead of vegetables from the field, hospitals instead of family care at home, driving a car instead of going by foot. It leads to the plundering of natural resources, swollen bureaucracies, and piles of garbage—which is precisely why modern industrial societies are on the verge of suffocating. Microelectronics, however, pledges to clean up the mess and, by offering the appropriate technological weapons, overcome industrialism's wretched legacy. The replacement of secretaries by word processors or computer surveillance of raw manufacturing inputs is finally nothing more than an expensive diet for society, intended to eliminate the excess weight of expansion. The fine-tuned society is the hidden ideal of the microelectronics revolution, a society in which all technological and social processes are controlled so that no friction arises and no waste is left behind.

While the computerized car promises to take care of poisons and pollution, the plans for electronic streets rest on the ambition of conquering congestion.

> Not only in the Federal Republic of Germany, but in all the European countries, the budgets of transportation ministries for new street construction have considerably diminished. But despite declining production and licensing figures for new motor vehicles, traffic planners anticipate an increase in new traffic demands. The space available for traffic will decline, traffic jams will become more common, and, as a consequence, the incidence of accidents will theoretically also rise. The main point in the future will therefore be to deal more rationally with the existing system of streets.[4]

4. *Frankfurter Rundschau,* January 29, 1983.

More vehicles in the same amount of space devoted to traffic—drivers will have to be able to assess traffic density, foresee traffic jams, decide on detours in advance, and select appropriate speeds. Naturally, their organic and intellectual capacities are not enough for the task; they can neither take in traffic conditions over a broad area like a bird, nor are they in a position to transform the multitude of variables quickly and conclusively into a strategy for proceeding. A computerized navigational system has both the overview and the processing capacity: sensors in or on the streets observe the traffic and transmit data to devices installed in automobiles, which compare traffic conditions to the destinations drivers have programmed and issue appropriate instructions via a video display. A commercial for Bosch makes the point:

> You want to travel [the twenty kilometers] from Recklinghausen to Dortmund. You get in your car and, before starting, type the destination number 1 140 980 (Dortmund) into your ALI on-board unit. The display tells you in which direction you have to drive, when to turn off, whether to anticipate ice, fog, or congestion. ALI leads you to your destination along the quickest route, because ALI knows what's going on throughout the traffic system. . . . With ALI, traffic flows more smoothly—drivers arrive more quickly and safely at their destinations, save energy, and protect the environment.

Traffic guidance systems are one element in the future projection of a fine-tuned society, in which supertechnology will halt the squandering of natural resources and time and the fraying of nerves. Is it not the well-situated in particular who are so strained by the demands of all kinds of goods and appliances, so overwhelmed by all the sales, appointments, and affairs requiring their attention, that they manage to make their rounds only in the greatest of haste? Here is where daily life starts going electronic: whether it is the "thinking washing machine," the cashless supermarket, or the electronic filing system, the aim of microelectronic innovations is to liberate people from organizational routines and information overload. The "intelligent highway" is just another example of the transformation of cultural aspirations in the age of the microchip. Whereas the automobile still lives off the promise of surpassing physical weakness through mechanical force, electronics promises to overcome psychological strain by expanding the horizons of information technology. It is in relief from stress—not, as before, relief from sweat—that the hope of information technology lies.

The internal combustion engine, that fundamental innovation that undertook to fight exhaustion and toil at the beginning of the century, has been surpassed not only technologically and economically, but culturally as well. To annihilate spatial distances by means of mechanical force—that is the motivation that has driven energy-intensive transportation from the first locomotive to the space shuttle. And it has now run up against its limits. A new cultural motivation has entered the scene with communications technology: to annihilate spatial distances by means of processing intelligence. Its attractiveness derives from adventures in mental space rather than adventures in geographical space. Now that the conquest of geographical space ends as often as not in traffic jams, the home computer comes along just in time to reinvigorate the enthusiasm automobile drivers have lost.

Less Is More

The old hopes cling tenaciously to life; a new generation of industrialists, bureaucrats, and engineers takes the stage to repair them. They put the finishing touches on engine concepts, experiment with fuels, install chips under the hood or in street pavements, or look to high-performance express trains for the realization of their dreams. In the name of a cleaner environment, industrialists capture new markets, bureaucrats take on new monitoring authorities, and engineers evaluate new career opportunities. They profit from the abuses of motorization and instill in the sobered public a fresh spirit for a technological boom. They have in mind a capital-intensive, administration-intensive, and research-intensive solution to the ecological crisis. Thus a new round in an old game begins: environmental degradation becomes a source of profit and prestige, as illness and criminality were in the earlier history of industrialization. There is no thought of dismantling the transportation compulsion or subduing the automobile. No, super-industrialization schemes call only for more ameliorative research and expanded sanitation industries. What once was to be had for nothing—clean air, stress-free mobility, safe streets, and peaceful sleep—will now be commercialized and produced in the form of special products and plans. Maintaining prosperity calls for added expenditures, and our reliance on a supply network, like our dependency in general, grows along with technological rearmament.

Two contrary visions of the potential course of society fuel the conflict over the automobile's future—and not only that of the automobile. Those gentlemen in gray flannel suits with Samsonite briefcases who populate the airports might be on their way to plan the high-tech extension of yesterday's cultural aspirations, while those taking off for bicycle rallies in their sneakers and neck scarves want to leave the obsessions of the nineteenth century behind and be moved by the idea of an unhurried society. Why should we, in the human equivalent of lemming behavior, go on seeking our future in "higher, faster, farther"? Is it not time to break the dictate of advancing motorization and bend automotive technology, like residential development, to the service of how we want to live? That does not mean that the automobile should be thrown onto the scrap heap of history; it probably does mean, however, that only a reformed traffic technology is capable of being combined with a respect for nature and the freedom to live independently of the motor. Today it is not culture that limps along behind technology, but the reverse. A transformation is on the agenda, one in which technology no longer stems from the dreamworld of the nineteenth century, but does justice to present longings for a society liberated from the compulsion toward progress.

Slower Speeds

As much as advertisements extolling economical and clean vehicles ring with the self-praise of automobile manufacturers, they hardly conceal that the true aim of the rationalization technicians is not to domesticate the automobile, but to make that monument to the love of speed, the speed machine itself, ever more streamlined through the addition of environmentally engineered accessories. Aerodynamic bodies, lightweight component parts, and electronic fuel injection are elements of a liturgical form that is meant to rescue the traditional creed in changed times: the high speed built into cars remains untouched, indeed, it achieves a new level after the efficiency cure. On close inspection, then, we find that the "rational automobile" has not yet come very far. Even supertechnology will fail to domesticate the automobile as long as cars are imaginable only as speed machines. Thus, for example—and despite ten years of efficiency rhetoric—the mean fuel consumption of 1970 vehicles was substantially lower than it is today, because the urge for bigger and higher-performance cars in individual class categories has eaten up the results of efficiency gains.

Actually, the path to a gentle car that conforms to its circumstances
is not so mysterious: we simply need prudently motorized vehicles.
Not speed machines for power mongers, but moderate engines for re-
laxed people. Decisive here is not more speed limit signs, radar traps,
and speeding tickets, but a self-regulation of speed that is built into
the engines themselves. Why should the blind urge for speed machines
be treated as a law of nature, when a reduction of engine performance
would open up new prospects for more humane traffic? Vehicles that
make do with a top speed of, say, ninety kilometers an hour and operate most efficiently at thirty to fifty kilometers an hour drastically reduce pollution without requiring one to give up the advantages of driving entirely. The gains would start with much lower fuel consumption,
because the vehicles would be driven more slowly and have smaller engines; go on to include less noise, because the tires would roll more
quietly, and see less wear, on streets intended for lower speeds; and
end with a radical decrease in traffic casualties, because, if not the number, certainly the severity of accidents drops with a decrease in overall
speed. New technological options would appear on the horizon—
might even electric cars become competitive?—and the automobile
would pose less of a danger to pedestrians and bicycle riders. The so-

cial expenditures involved in automobile traffic would fall as well, because cars driven at slower speeds would necessitate fewer insurance agencies, court sessions, and police patrols. The way out of the dilemma between sinking utilities and rising burdens is found in a clever limitation on technological excess. The speed machine, ultimately, has no place in a society hospitable to nature and human beings.

Today's fleet of automobiles is grotesquely overpowered, with all the consequent waste of energy, high materials consumption, and loss of security. Acceleration capabilities and top speeds are treated as if cars had to withstand long-distance races every day. Yet on average, cars spend 80 percent of their operating time in city traffic, at an average speed of twenty-five kilometers an hour! In actual use, then, automobiles are a means more of short- than of long-distance travel, and they need to be capable of no more than a cozy cruising speed: to send speed machines out onto the streets is about as rational as shooting at sparrows with cannons. High-performance motorization has as much to do with transportation requirements as Gothic cathedrals did with shelter from inclement weather. Speed machines are the material expression of the religion of progress, of the future-consuming society that sacrifices today to tomorrow with scarcely a thought.

For an unhurried society, however, everything depends on developing an intensive and attentive way of treating the present. Such a society does not need to rush headlong into the future, for it is not driven by the fear that it will miss something. In a culture of calm composure, athletic, triumphant drivers make a ridiculous impression, because they betray to watching eyes their feeling of being pursued by a deficit. Nineteenth- century society was driven to haste because it feared backwardness; a self-confident society of the twenty-first century will once again be able to afford slower speeds.

Shorter Routes

The outfitting of cars with top-notch technology offers no protection—rather, the contrary—from the environmental crisis of the second order now occurring in total motorization: namely, the erosion of nearby spaces that beckon one to a nonmotorized lifestyle. Distances grow, life comes unraveled in space, and neighborhoods degenerate into transit routes: the environment favorable to pedestrians and bicyclists atrophies.

In the future designs of an easygoing society, speed—or lack of it—

plays a key role. Distances grew because it became possible to travel faster; why should distances not shrink again once people start traveling more slowly? A low-speed society would suggest the desirability of shorter routes and contribute to the braking of the ominous development that made the distant overtake the near in the first place—who knows, perhaps development could even be rolled back a notch. Clearly, though, the near must be revitalized if the compulsion toward the automobile is ever to be dismantled.

To help the near reassert its right over the distant means first of all to create zones liberated from speed. Children could play hopscotch in the street and old men and women could resume their chats under the trees once the dictatorship of penetrability is undermined and the streets are no longer tailored to express traffic. Bicyclists and pedestrians could breathe easily and would again have something to experience on the street once cars are required to make their way with caution. Even trees, benches, and fountains could renew their claim to space. It would be worthwhile to use one's legs again once shops, bars, and businesses reenter the neighborhood.

Recent efforts at traffic mitigation represent the tentative beginnings toward relieving the automobile of its monopoly on the streets and facilitating a peaceful coexistence between motorized and nonmotorized traffic. A speed limit of thirty kilometers per hour in cities would also lessen the absurd competitive pressure on public transit—visible in monstrous subway constructions—to match the pace of the automobile. The logical continuation of traffic mitigation would be the large-scale, ecological transformation of city districts, in which traffic surfaces would be reclassified, planted with grass, and returned to the population for multifarious uses. This would allow an economic structure devoted to detail and nuance to become established, one whose concentration of goods and services would facilitate the development of a daily lifestyle centered on short journeys.

Such a reconstruction of the immediate vicinity breaks with a basic dogma of the restless society: that progress means minimizing the resistance of distance and making space more penetrable. The right to visit a distant place now retreats behind the right to recapture one's own place; the habitability of the immediate vicinity will no longer be sacrificed to the accessibility of distant locales.

This new appreciation of nearness does not arise by accident. It is inherent in an image of a locally centered economy that no longer takes pride first of all in faraway exports, but in the density of nearby exchanges that increase independence. Those who see for the future a tightly interwoven national and international society cannot give up speed and penetrability. Those, however, who hope for an economy that is locally denser and internationally less enmeshed can afford prudence and insist on the integrity of the near.

Policy for the New Freedom

Slower speeds and shorter routes are the cornerstones of a policy that aims at dismantling the political and economic assumptions of society based on the automobile. The project of total, high-performance motorization, namely, can survive only if nature's right to life and the freedom of movement of half the population are sacrificed to it. Conversely, respect for nature and the right of independent, nonmotorized locomotion can be preserved only if both speed and distance are returned to a moderate level. Beyond a certain threshold of industrial growth, an increase in the quality of life will come only by limiting output, for only then can space for noneconomic activity be recreated. In the case of transportation, latitude for new forms is created not by blind breakthroughs into the future, but by sovereign self-limitation. Slower speeds and shorter routes could lead to a new flourishing in the quality of life, because they leave nature intact and help autonomous movement attain a new power.

In particular, a policy of ease depends on gradually spreading the idea of being able to choose a life without a car without giving up too many advantages. Slower speeds and shorter routes are the preconditions for a new right in freedom of movement—namely, the right not to be discriminated against for not possessing an automobile. This right not only protects nonmotorized mobility, but also secures a high level of accessibility for the majority that either cannot or does not want to take the wheel. Indeed, this freedom of movement promises even more, for it keeps open the option of living with less money but still pleasantly.

If a less income- and consumption-intensive society is to be effected—but one that does not entail a loss in quality of life—compulsive consumption must be done away with. Living conditions that do not, for better or worse, compel one to depend on an automobile keep the future of a posteconomic society open. The Study Group on Traffic of the Bundesverband Bürgerinitiativen Umweltschutz (Federal Environmental Initiative Association) formulated, not without irony,

Photograph by Robert Rodale.

the maxims of such a policy based on the federal constitution of West Germany:

Article 3 (Equality Before the Law)
    1.   Everyone is also equal before the traffic regulations.
    2.   Pedestrians, bicyclists, users of public transportation, and automobile drivers have equal rights.
    3.   No one may be disadvantaged or accorded advantages on account of his or her chosen means of transportation, speed, or physical condition.

Article 4 (Freedom of Choice in Transportation)
    1.   The freedom of choice in means of transportation is to be secured.
    2.   No one may be compelled, by structural, legal, or other measures, to the use of an automobile.

Article 5 (Freedom of Travel)
    1.   All Germans enjoy freedom of travel in all areas of the Federal Republic.
    2.   No one may be compelled, for reasons of the exercise of profession, shopping, leisure activities, or physical or emotional recuperation, to cover long distances.[1]

1. *Informationsdienst Verkehr* (Traffic information service), ed. Arbeitskreis Verkehr im BBU (Working Group on Traffic, Federal Environmental Initiative Association), Berlin, no. 12 (July 1983).

Selected Bibliography

Allensbacher Jahrbuch für Demoskopie. Allensbach, 1977.

Anders, Günter. *Die Antiquiertheit des Menschen,* vol. 2. Munich, 1980.

Attali, Jacques. *Histoires du temps.* Paris, 1982.

Bade, Wilfried. *Das Auto erobert die Welt: Biographie des Kraftwagens.* Berlin, 1938.

Bataille, Georges. *The Accursed Share: An Essay on General Economy.* Trans. Robert Hurley. New York, 1988.

Baudelot, Christian, Roger Establet, and Jacques Toiser. *Qui travaille pour qui?* Paris, 1979.

Baundry de Saunier, L. *Grundbegriffe des Automobilismus.* Vienna, 1902.

Beduhn, Ralf. *Die Roten Radler: Illustrierte Geschichte des Arbeiterrad-fahrerbundes "Solidarität."* Münster, 1983.

Berger, H. J., G. Bliersbach, and R. G. Dellen. *Macht und Ohnmacht auf der Autobahn: Dimensionen des Erlebens beim Autofahren.* Frankfurt, 1973.

Berger, Michael D. *The Devil Wagon in God's Country: The Automobile and Social Change in Rural America, 1893–1929.* Hamden, Conn., 1979.

Betz, Ludwig. *Das Volksauto: Rettung oder Untergang der deutschen Auto-mobilindustrie.* Stuttgart, 1931.

Bierbaum, Otto Julius. *Eine empfindsame Reise im Automobil: Von Berlin nach Sorrent und zurück an den Rhein.* Munich, 1903.

———. *Yankeedoodlefahrt und andere Reisegeschichten.* Munich, 1910.

Bremer, Robert. *So leben wir morgen: Der Roman unserer Zukunft.* Munich, 1972.

Burckhardt, Lucius. "Landschaftsentwicklung und Gesellschaftsstruktur." In

F. Achleitner, ed., *Die Ware Landschaft: Eine kritische Analyse des Land-schaftsbegriffs*, 9–15. Salzburg, 1977.

Callenbach, Ernest. *Ecotopia: The Notebooks and Reports of William Weston.* Berkeley, 1975.

Ceron, Jean-Paul, and Jean Baillon. *La société de l'ephémère.* Grenoble, 1979.

Dahrendorf, Ralf. *Gesellschaft und Demokratie in Deutschland.* Munich, 1965.

Daumier, Honoré. *Les transports en commun.* Paris, 1976.

Dettelbach, Cynthia D. *In the Driver's Seat: The Automobile in American Literature and Popular Culture.* Westport, Conn., 1976.

Dieckmann, Hans. "Das Auto als Traumsymbol." *Analytische Psychologie* 7 (1976): 20–35.

Diesel, Eugen. *Wir und das Auto: Denkmal einer Maschine.* Leipzig, 1933.

———. *Autoreise 1905.* Leipzig, 1941.

Douglas, Mary, and Jason Isherwood. *The World of Goods: Towards an Anthropology of Consumption.* New York, 1978.

Eckermann, Erik. *Von Dampfwagen zum Auto.* Reinbek, 1981.

Eichberg, Henning. *Leistung, Spannung, Geschwindigkeit: Sport und Tanz im gesellschaftlichen Wandel des 18./19. Jahrhunderts.* Stuttgart, 1978.

———. "Die 'Revolution des Automobils.' Materialien zu einer kritischen Technikhistorie, die das Verschwinden der Sachen mitdenkt." In *Die historische Relativität der Sachen.* Münster, 1984.

Ewen, Stuart. *Captains of Consciousness: Advertising and the Social Roots of the Consumer Culture.* New York, 1976.

Fleischer, Arnulf. "Zur Modell- und Innovationspolitik der Automobilhersteller." In Gerhard Bodenstein and Helmut Leuer, eds., *Geplanter Verschleiß in der Marktwirtschaft*, 466–505. Frankfurt, 1977.

Flink, James J. *The Car Culture.* Cambridge, Mass., 1976.

Francke, Peter, ed. *Lob des Fahrrads.* Zurich, 1974.

Frankenberg, Richard von. *Geschichte des Automobils.* Künzelsau, 1973.

Garbrecht, Dietrich. *Gehen: Plädoyer für das Leben in der Stadt.* Weinheim, 1981.

Glaser, Hermann. *Maschinenwelt und Alltagsleben: Industriekultur in Deutschland von Biedermeier bis zur Weimarer Republik.* Frankfurt, 1981.

Grauhan, Rolf-Richard, and Wolf Linder. *Politik der Verstädterung.* Frankfurt, 1974.

Grube, Frank, and Gerhard Richter. *Das Wirtschaftswunder: Unser Weg in den Wohlstand.* Hamburg, 1983.

Harmelle, Claude. *Les Piqués de l'aigle: Saint Antonin et sa région (1850–1940). Révolutions des transports et changement social.* Paris, 1982.

Heinze, Wolfgang G. "Verkehr schafft Verkehr." *Berichte zur Raumforschung und Raumplanung* 23, nos. 4–5 (1979).

Henning, Hansjoachim. "Kraftfahrzeugindustrie und Autobahnbau in der Wirtschaftspolitik des Nationalsozialismus, 1933–1936." *Vierteljahresschrift für Sozial- und Wirtschaftsgeschichte* 65 (1978): 217–42.

Hickethier, K., W. D. Lützen, and K. Reiss. *Das deutsche Auto: Volkswagen Werbung und Volkskultur.* Steinbach, 1974.

Hirsch, Fred. *Social Limits to Growth*. Cambridge, Mass., 1976.

Holzapfel, Helmut, Klaus Traube, and Otto Ullrich. *Auto-Verkehr 2000: Wege zu einem ökologischen und sozialverträglichen Automobilverkehr*. Karlsruhe, 1985.

Illich, Ivan. *Energy and Equity*. London, 1974.

Jungwirt, Nikolaus, and Gerhard Kromschröder. *Die Pubertät der Republik*. Frankfurt, 1978.

Kaftan, Kurt. *Der Kampf um die Autobahnen: Geschichte und Entwicklung des Autobahnsgedankens in Deutschland von 1907–1935 under besonderer Berücksichtigung ähnlicher Pläne und Bestrebungen im übrigen Europa*. West Berlin, 1955.

König, René. *Kleider und Leute: Zur Soziologie der Mode*. Frankfurt, 1967.

Korte, J. W., ed. *Stadtverkehr: gestern, heute und morgen*. Berlin, 1959.

Krämer-Badoni, Thomas, Herbert Grymer, and Marianne Rodenstein. *Zur sozio-ökonomische Bedeutung des Automobils*. Frankfurt, 1971.

Kuke, Paul. "Hitler und das Volkswagenprojekt." *Vierteljahresheft für Zeitgeschichte* 8 (1960): 341–83.

Kutter, Eckhard. "Mobilität als Determinante städtischer Lebensqualität." In Wilhelm Lentzbach, ed., *Verkehr in Ballungsräume*. Cologne, 1975.

Lärmer, Karl. *Autobahnbau in Deutschland 1933–45: Zu den Hintergründen*. Berlin, 1975.

Lartique, Jacques H. *Photo-Tagebuch unseres Jahrhunderts*. Lucerne, 1970.

Lay, Lothar, ed. *Bruno Paul*. Munich, 1974.

Leiss, William. *The Limits to Satisfaction: An Essay on Needs and Commodities*. Toronto, 1976.

Lessing, Hans-Erhard, ed. *Fahrradkultur 1. Der Höhepunkt um 1900*. Reinbek, 1982.

Linder, Staffan B. *The Harried Leisure Class*. New York, 1970.

Lötschburg, Winfried. *Von Reiselust und Reiseleid: Eine Kulturgeschichte*. Leipzig, 1977.

McLellan, John. *Bodies Beautiful: A History of Car Styling and Craftsmanship*. London, 1975.

Mai, Ekkehard. "Das Auto in Kunst und Kunstgeschichte." In Tilmann Buddensieg and Bernd Rogge, *Die nützlichen Künste*. Berlin, 1980.

Maissen, Felici. *Der Kampf um das Auto in Graubünden, 1900–1925*. Chur, 1968.

Mander, Helmut. *Automobilindustrie und Autosport: Die Funktionen des Automobilsports für den technischen Fortschritt, für Ökonomie und Marketing von 1894 bis zur Gegenwart*. Frankfurt, 1978.

Mellinghoff, Hans. *Aufbruch in das mobile Jahrhundert: Verkehrsmittel auf Plakaten*. Dortmund, 1981.

Mende, Hans Ulrich von. *Styling: automobiles Design*. Stuttgart, 1979.

Mumford, Lewis. *The Myth of the Machine: The Pentagon of Power*. New York, 1970.

Nash, Dennison. "The Rise and Fall of an Aristocratic Tourist Culture, Nice 1763–1936." *Annals of Tourism Research* 6 (1979): 61–75.

Nelson, Walter H. *Small Wonder: The Amazing Story of the Volkswagen.* Boston, 1965.

Overy, J. "Cars, Roads and Economic Recovery in Germany, 1932–38." *Economic History Review* 28 (1975): 466–83.

Petsch, Joachim. *Geschichte des Auto-Design.* Cologne, 1982.

Pichois, Claude. *Vitesse et vision du monde.* Neuchâtel, 1973.

Pidoll, Michael, Freiherr von. *Der heutige Automobilismus: Ein Protest und Weckruf.* Vienna, 1912.

Plowden, Stephen. *Taming Traffic.* London, 1980.

Plowden, William. *The Motor Car and Politics, 1986–1970.* London, 1971.

Polster, Bernd. *Tankstelle: Die Benzingeschichte.* Berlin, 1982.

Poulain, Hervé. *L'art et l'automobile.* Zug, 1973.

Rauck, Max, Gerd Volke, and Felix R. Paturi. *Mit dem Rad durch zwei Jahrhunderte: Das Fahrrad und seine Geschichte.* Aarau, 1979.

Rennert, Jack. *100 Jahre Fahrradplakate.* Berlin, 1974.

Reulecke, Jürgen. "Vom blauen Montag zum Arbeiterurlaub: Vorgeschichte und Entstehung des Erholungsurlaubs für Arbeiter vor dem Ersten Weltkrieg." *Archiv für Sozialgeschichte* 16 (1976): 205–48.

Riedel, Manfred. "Vom Biedermeier zum Maschinenzeitalter. Zur Kulturgeschichte der ersten Eisenbahnen in Deutschland." *Archive für Kulturgeschichte* 43 (1961): 100–123.

Robert, Jean. *Le temps qu'on nous vole: Contre la société chronophage.* Paris, 1980.

Rothschild, Emma. *Paradise Lost: The Decline of the Auto-Industrial Age.* New York, 1974.

Sahlins, Marshall. *Culture and Practical Reason.* Chicago, 1976.

Saka, Pierre, Jean Menu, and Jean-Pierre Dauliac. *Histoire de l'automobile en France.* Paris, 1982.

Schaeffer, K. H., and Elliott Sclar. *Access for All: Transportation and Urban Growth.* Harmondsworth, 1975.

Schivelbusch, Wolfgang. *The Railway Journey: The Industrialization of Time and Space in the Nineteenth Century.* Berkeley, 1986.

Scitovsky, Tibor. *The Joyless Economy: An Inquiry into Human Satisfaction and Consumer Dissatisfaction.* New York, 1976.

Sedgwick, Michael. *Die schönsten Autos der 30er und 40er Jahre.* Dusseldorf, 1982.

———. *Die schönsten Autos der 50er and 60er Jahre.* Dusseldorf, 1983.

Seherr-Thoss, H. Chr. Graf von. *Die deutsche Automobilindustrie: Eine Dokumentation von 1887 bis heute.* Stuttgart, 1974.

Seper, Hans. *Damals als die Pferde scheuten: Die Geschichte der österreichischen Kraftfahrt.* Vienna, 1968.

Siebertz, Paul. *Carl Benz: Ein Pioneer der Verkehrsmotorisierung.* Munich, 1943.

Sombart, Werner. *Luxus und Kapitalismus.* Berlin, 1922.

Spörli, Siro. *Seele auf Rädern: Psychologie auf der Straße.* Olten, 1972.

Stommer, Rainer, ed. *Reichsautobahn: Pyramiden des Dritten Reichs.* Marburg, 1982.

Stone, Norman. *Europe Transformed, 1878–1919*. London, 1983.

Voswinckel, Peter. *Arzt und Auto: Das Auto und seine Welt im Spiegel des Deutschen Ärtzeblattes von 1907 bis 1975*. Münster, 1981.

Wachtel, Joachim. *Facsimile: Querschnitt durch frühe Automobilzeitschriften*. Munich, 1970.

Wendorff, Rudolf. *Zeit und Kultur: Geschichte des Zeitbewußtseins in Europa*. Opladen, 1980.

Wettich, Hans. *Die Maschine in der Karikatur: Ein Buch zum Siege der Technik*. Berlin, 1916.

Wik, Reynold M. *Henry Ford and Grass-roots America*. Ann Arbor, 1972.

Williams, Raymond. *Keywords: A Vocabulary of Culture and Society*. London, 1976.

Wolf, Wilhelm. *Fahrrad und Radfahrer*. Leipzig, 1980.

Wolff, Theo. *Vom Ochsenwagen zum Automobil: Geschichte der Wagenfahrzeuge und des Fahrwesens von ältester bis neuester Zeit*. Leipzig, 1909.